# THE SOUTH FLORIDA

# GARDENING
# SURVIVAL GUIDE

# THE SOUTH FLORIDA
# GARDENING SURVIVAL GUIDE

## by David The Good
### Illustrated by Tom Sensible

GOOD
BOOKS

*The South Florida Gardening Survival Guide*

David The Good

Copyright © 2022 by David The Good

Illustrated by Tom Sensible

Good Books Publishing

goodbookspub.com

ISBN: 978-1-955289-11-5

# Contents

# Introduction

When I was about eight years old, I visited my great-grandpa in Upstate New York and saw his incredible garden. It was a long way from my parents' home in sunny Ft. Lauderdale. Great-grandpa's garden had long rows of potatoes and beets, many different green vegetables and long hedges of raspberries and blackberries. In his front yard giant sunflowers were scattered among the landscaping. It was incredible! His garden looked like the classic country garden you imagine from the 1940s; nice, neat rows of beautiful crops growing in dark, loamy soil.

"Here David, take some of these," he said, pressing a small brown paper packet of seeds into my hand. "Plant these beets in your garden. And here," he continued, giving me a little bag of white powder. "This is lime. Put it in the garden when you plant."

When I got home, I planted my garden and tried to grow like he did. Yet I had no luck, failing to reap even a single beet.

Year after year I tried gardening, reading the backs of the seed packets and doing my best, but my yields were marginal and most of what I planted died or failed to thrive. It seemed like nothing liked the hot sand and the heat.

As I got older, however, I did discover some things that would grow without much effort on my part. Snake beans from Southeast Asia, nopale cactus from Mexico, yams from Africa and other oddities from around the world. They weren't the sorts of things that my Great-grandpa grew, but they liked South Florida and bore me food, unlike those poor, sad beets! What an encouragement it was to find vegetables that not only survived but yielded wonderfully in our climate.

After decades of Florida gardening, I am a pro at working with our soil. I wrote the popular books *Totally Crazy Easy Florida Gardening, Create Your Own Florida Food Forest,* and *Florida Survival Gardening* to help others find the same success I've found. Along the way, I also learned how to grow beets and other cold-climate vegetables too, and I will share some tips on those along with my recommendations for some of the Florida-friendly tropical vegetables I use as the core of my gardens.

There are multiple books on gardening in Florida, yet Florida is a big state, ranging from tropical at the tip to temperate at the top. There are four entire USDA growing zones in just one state, and covering all that space means losing a little bit of laser focus on what each area specifically needs to thrive.

A frustrated gardener asked me in 2020 if I would consider

writing a book just for gardeners living in South Florida. The idea appealed to me, as I absolutely love the climate of the city where I grew up and am happy to focus on making South Florida backyards into food-producing oases. Once I finished my massively expanded re-write of *Create Your Own Florida Food Forest*, it was time to get this little book written and published.

In this book you'll discover the joy of South Florida gardening and the abundance of exciting plants you can grow and eat. You'll learn to embrace the sand and the rain and the heat and turn your backyard into something beautiful. You'll also save a bundle by growing a lot of good food in even a small space. I'll cover some of your favorite vegetables along with some new ones. And we may even figure out how to beat the iguanas.

And on that note, in this book I am very pleased to be able to feature the entertaining art of my friend and fellow gardener Tom Sensible. His iguanas are the only iguanas I would ever want on my property.

But enough of the introduction—let's jump into South Florida gardening!

CHAPTER 1:

# The Challenges of South Florida Gardening

**M**any transplants to South Florida see it as a gardening wasteland. "It's SO HARD to garden here!" we cry as we sweat in the December heat. "Back in Newyormich-illinerseysotaland we had real soil, and the freezes killed all the bugs, you know. My Grandma grew tomatoes the size of your head, and she'd can truckloads of them. You can't grow a thing in this sand. I tried planting some Brandywines this last April and they didn't even come up!"

No, it ain't easy to grow them dang Yankee crops. There are a lot of challenges in South Florida for gardeners, especially new ones. Starting with the "soil."

## The Sand is Coarse and Irritating and It Gets Everywhere

Seriously, what is this stuff? It's not soil. You can watch a lot of YouTube gardening videos without seeing a single garden that has anything like the stuff we have in South Florida. My Ft. Lauderdale food forest garden has gray sand that makes your hands and feet dirty. Dig a little deeper and you get pure white sand that looks like it's made for sandboxes or mixing concrete. As a kid I mowed the next-door neighbor's "grass" (it was more of a patch of struggling weeds) and the lawn-mower would kick up clouds of dust that smelled like burned rocks. Some people say it's like beach sand, but it isn't. Beach sand is pleasant to play with, whereas the white-gray sand isn't that nice. It drains really well, though, and has coarse grains which means that you can do cool tricks like sticking the end of a running hose into the ground and pushing it deeper and deeper into the earth until the entire hose is buried somewhere in the earth's core. Try it, it's fun! You won't get the hose back again and will have to cut it off at ground level, but it's sure an entertaining way to spend ten minutes of time. Unless you and your best friend are trying this trick in his backyard with a brand-new hose and his dad catches you and gets really, really mad. Don't ask.

Despite the great drainage, sometimes Fort Lauderdale sand won't accept water at all. When the ground is dry and you run the hose, the water just runs off in every direction and refuses to soak in. This condition is called being "hydropho-bic," and is a result of surface tension not letting the water go through. It's really strange when you're used to normal soil.

If you decide you want to build a nice garden by adding

compost and mulching, you'll find that it takes a lot of humus to turn the sand into something that looks like soil. You'll also find that the sand EATS compost at a startling rate. I added *tons* of compost, purchased soil, peat moss and mulch to my gardens over the years and that Florida sand ate it all. You'd never know I'd done a thing to it. Eventually the dead grey sand wins.

There are places in South Florida where the soil looks better, and that's usually under the canopy of live oaks or other trees where the ground is moist and shaded and receives regular leaf drops. Yet if you clear the area and put in an annual garden, it reverts quickly to hot, grey sand.

Put in drip irrigation and the drips soak almost straight down into the ground rather than spreading out nicely. Throw down 10-0-10 and it washes through with the first good rain. Till in some compost and it's like throwing a delicious corn-fed rat into a piranha tank. Boom—gone!

Yes, oh tender transplant, the sand is a difficulty!

## The Heat isn't just a Basketball Team

On the back of your packet of Yankee pumpkin seeds it says "plant in spring, after all danger of frost has passed." You read it in May and think, "hey, it hasn't frozen here in 15 years ... guess I can plant now!" So you do. The seeds come up and grow okay for a bit, then the rains of June start pounding, interspersed with sauna heat and humidity and next thing you know, all the pumpkin vines are sick and moldy and wilting at noon every day. And then they die without bearing a single pretty Jack-O-Lantern like the one on the front of the packet being held by some cute child in some weirdly perfect garden up North.

It's so sad. The time of year when the rain comes in is also the hottest and least hospitable to "normal" garden crops, like tomatoes and bell peppers, lettuce, cabbages, and radishes. They just won't grow much—if at all—in the summer heat, no matter what you read on a seed packet.

## Petty Tyrants

South Florida has more than its fair share of tyrants that don't want you to have any control over your own land. If you plant a garden in the "wrong" spot, you might face the wrath of code enforcement, the condo commandos, the city Karens and who knows what else. You can barely sneeze without a permit. I once gave a talk in a planned community and was horrified to discover that almost every edible plant I recommended was not on the "allowed" list in that neighborhood. "I wish we could grow those," one woman said, looking at some of the fruit trees I'd brought with me.

Yet some of this is changing. You may get some push-back from the neighborhood if you put in a front-yard garden, but as of 2019, there is an excellent new law on the books in Florida.

*(1)    The Legislature intends to encourage the development of sustainable cultivation of vegetables and fruits at all levels of*

*production, including for personal consumption, as an impor-*
*tant interest of the state.*

*(2)     Except as otherwise provided by law, a county, municipality,*
*or other political subdivision of this state may not regulate vegeta-*
*ble gardens on residential properties. Any such local ordinance or*
*regulation regulating vegetable gardens on residential properties*
*is void and unenforceable.*

*(3)     This section does not preclude the adoption of a local ordi-*
*nance or regulation of a general nature that does not specifically*
*regulate vegetable gardens, including, but not limited to, regula-*
*tions and ordinances relating to water use during drought condi-*
*tions, fertilizer use, or control of invasive species.*

*(4)     As used in this section, the term "vegetable garden" means a*
*plot of ground where herbs, fruits, flowers, or vegetables are culti-*
*vated for human ingestion.*

This looks to be pretty wide open and may help the many gardeners in the state who have had to fight to keep their gardens, sometimes losing their plots of vegetables—and, in one case I am familiar with, the entire property and house!—due to restrictive laws.

You're still going to have to fight in some cases, because there are lots of picky, picky people in South Florida. Neighbors complain about the "mess" of fruit, or banana trees attracting roaches, or raised beds looking ugly, and all kinds of nonsense, so keep that in mind. As much as you try to be friendly with neighbors, there are rotten people out there that just won't be happy unless they control your life.

## The Pests Never Die

In South Florida the pests are immortal, feeding voraciously on the dreams of gardeners. There are no frosts to end the plagues, and no Moses to stretch out his staff and call them off no matter how you repent of your gardening follies. You'll meet weird beetles and caterpillars, fire ants and leaf-footed bugs, stink-bugs and hornworms, mole crickets and chinch bugs, mealy bugs and creepy white weevils—and if your garden survives all of those, there is always an iguana ready to munch his way through the salad bar you have so graciously provided.

In Newyormichillinerseysotaland, there are freezes that knock the pests back to a reasonable level every winter. Not in South Florida. Once you start gardening, they'll keep showing up and growing until they're powerful enough to devour everything you love.

Looking at the above list, are you sad? I am. I guess we just need to skip gardening in South Florida and do something else, like drink smoothies and watch TLC. Yeah, I might as well just leave the rest of this book blank. I'm sorry I even brought it up. So—what TV shows do you like?

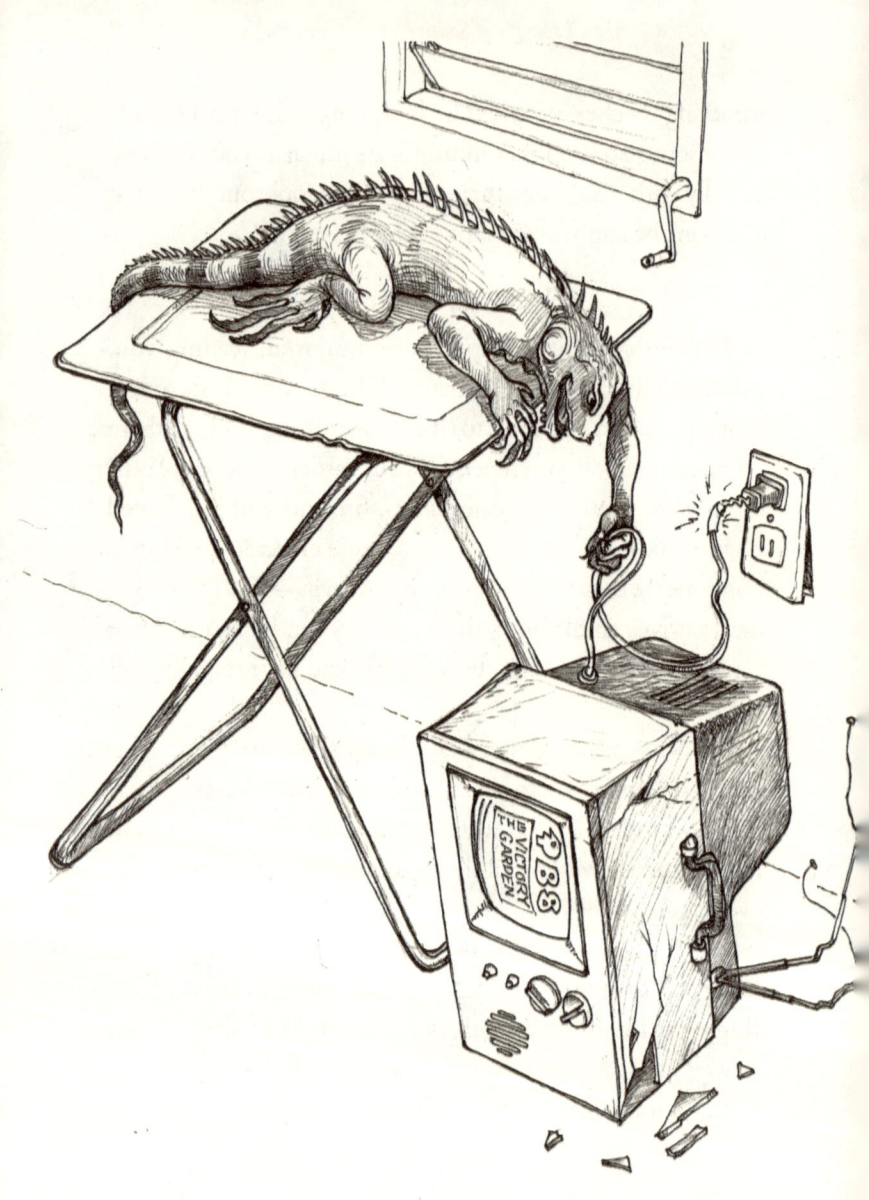

CHAPTER 2:

# The Wonders of South Florida Gardening

Ha! You thought I was going to quit? I hate TLC since Honey BooBoo went off the air. And I'm allergic to the emulsifiers in smoothies. And an iguana broke my TV anyhow.

Let's talk about the good side of gardening in South Florida. The Glorious, Incredible, Amazing Wonders of South Florida Gardening!

### The Sand is Easy to Dig

Yes, I know sand is coarse and irritating, but it could be worse. Sure, it eats compost and all that, but planting stuff is easy. So is pulling weeds. You can grab the grass and yank it out with your bare hands in clumps. Want to make a raised bed? Rake up a mound in five minutes and you've got one. Want to put in a fence post? You can dig the hole with your kid's plastic beach shovel. I've lived in places with rocks and clay and I'd take Florida's sand over any of them. Sure, it's almost nutrient free, but it's easy. You don't have to worry about waterlogging or tilling in the wrong season or hardpan. Sand is very forgiving and convenient. Sand is the microwavable burrito of soils. Not the best on the nutrition scale, but certainly easy.

## Good Rainfall

South Florida has good rainfall. Yes, we have our dry months, but they're also cool. When the heat really kicks in, the rain comes with it. It's wonderful when the breeze and rain start up in the middle of a hot day, drenching you as you work in the garden. My favorite are the days when the sun retreats and everything turns strangely green as towering blue-black thunderheads sweep in and cool gusts of rain-scented wind whip through the palm fronds. Often a storm blows through in just a few minutes, then the tropical sun returns and raises steam from the wet blacktop roads. It's beautiful.

## Year-Round Gardening

Up north people must stop gardening for a big chunk of the year. Can you imagine that? How sad! In South Florida you never have to stop. Ever. In the winter you can grow cool-season vegetables and in the summer you can tend your tropical fruits and vegetables. There's never a time of year where you have to wrap yourself in blankets while looking longingly through seed catalogs. Because of this year-round season, you can actually produce a ton more food—literally—in a South Florida yard than you could in a bigger yard up north. Think about it: if your frost-free growing season up north was about 120 days, then you get the equivalent of *three* growing seasons in South Florida, as we have 365 days of frost-free weather. 1/3 acre in a year-round growing season can produce the same amount as a full acre in a climate with a short growing season. It gets better than that, though, because South Florida has ...

## The Riches of the Tropics on Tap!

Freezes kill many of the world's best food plants. The farther north you go, the fewer species you can grow. At the equator there are thousands of varieties of fruit. Up north there are only dozens. Sure, we're familiar with that dozen, since many of us have European ancestry, so they seem like everything. Apples, pears, raspberries, plums, currants, blueberries ... many of the familiar fruits we enjoy are not South Florida friendly. The few that we know are oranges, bananas, avocados, and pineapples. But the tropics have much more than that! When you reset your focus, you'll see that South Floridians are much richer in garden variety than northerners. Mangoes, vanilla, coffee, jackfruit, loquats, Jamaican cherry, mangoes, wax apples, mamey sapote, chocolate pudding fruit, ice cream bean, allspice, starfruit, key lime, mangoes, plantains, Surinam cherry, tamarind, acerola cherry, guava, mangoes, jabuticaba, soursop, sugar apple, Governor's plum, mangoes, Mysore raspberry ... the list goes on and on and on. And did I mention mangoes? I love mangoes. They can grow to the size of oaks in South Florida.

## Orchids on Trees

You can grow orchids outside in South Florida. Just stick them on your trees and they grow. I know, they're aren't edible, but it's still amazing. A family friend from Minnesota grew some beautiful orchids in my mom's backyard while she was working on her medical degree in South Florida. When she left—without taking her plant collection—the orchids just kept growing. Right now there are a half-dozen of them blooming without care around Mom's backyard.

## South Florida has lots of Plant People

There are great plant nurseries around the Southern portion of the state that grow plants which won't grow anywhere else in the continental United States (with the possible exception of Southern California). There are gardening clubs and botanical gardens of all sorts and lots and lots of private collections of plants that will blow your mind. The sheer diversity is amazing. Take pineapples as an example. There are large juicing pineapples and small dessert pineapples, brilliant pink pineapples and dark green pineapples. There are red pineapples and white pineapples, and pineapples with variegated leaves. You could design a landscape bed with just different varieties of pineapples. And that's just pineapples!

People collect and share fruit and vegetable varieties all the time, and nurseries propagate and sell all kinds of wonderful things. Make friends and go shopping—you'll be amazed by what you can find. Right off the top of my head, I recommend checking out ECHO Global Farm in Ft. Myers, The Fruit & Spice Park in Homestead and the Rare Fruit and Vegetable Council of Broward County. All will inspire you.

But that's enough about the wonders of South Florida gardening—let's get into the "how to," so we can make your gardening dreams come to life.

CHAPTER 3:

# How to Grow a Great South Florida Vegetable Garden

G rowing vegetables is not impossible in South Florida, no matter what some may say. You can do it; you just need to adjust how and what you grow. As any gardener will tell you, healthy plants start with healthy soil. But as we know in South Florida, THERE IS NO SOIL! What's a gardener supposed to do? Here are some options for fixing and managing what you have.

## So, Can We Fix the Crummy Dirt?

As mentioned previously, South Florida sand burns through organic material. And in some spots, you just have limerock without even sand. That's a serious problem. Most gardening books have nice little sections on building soil with compost and manure and sheet mulching and all that, and it looks so pretty and nice, but Florida doesn't play nice. It likes to eat your amendments. When I built The Great South Florida Food Forest Project some years back, I stacked roughly three feet of mulching materials on the ground, including hunks of logs, chicken manure, palm fronds, yard waste and every other bit of cover we could find. It turned into compost quickly, and then that compost disappeared into the sand. There are sandy patches all over the food forest now, despite repeated applications of organic material. To maintain a semblance of soil in South Florida takes effort. Leave it alone for a year or so and you're back to sand. This is just life here. Gardeners must keep plenty of organic material on hand and feed the ground regularly. It's just part and parcel of living in paradise. I've done it for years.

Though I may have rediscovered a way to improve the sand for good, turning it into a Floridian equivalent of "terra preta." What is terra preta? It's the "black earth" of the Amazon.

Like South Florida, the Amazon rainforest generally has very poor soil, yet there are pockets of fertile ground in the Amazon that differ from the poor soil around them. Called "terra preta," this dark, carbon-rich soil has retained its fertility for centuries, but its origin is shrouded in mystery.

As Johannes Lehmann at Cornell University notes:

**"Already at the end of the 19th century, Smith (1879) and Hartt (1885) reported the existence of dark earths in the Amazon, which had a dark color and were**

**highly fertile. The origin of the Amazonian Dark Earths is not entirely clear and several conflicting theories were discussed in the past. Camargo (1941) speculated that these soils might have formed on fallout from volcanoes in the Andes, since they were only found on the highest spots in the landscape. Other theories included a formation as a result of sedimentation in Tertiary lakes (Falesi, 1974) or in recent ponds (Cunha-Franco, 1962). Further theories are mentioned by Smith (1980), which all did not hold against later investigations. It is now widely accepted that these soils were not only used by the local population but are a product of indigenous soil management as proposed by Gourou (1949). Later surveys confirmed these findings (Sombroek, 1966; Smith, 1980; Kern and Kämpf, 1989). Whether they were intentionally created for soil improvement or whether they are a by-product of habitation is not clear at present. This is in part due to the varied features of the dark earths throughout the Amazon Basin."**

There are patches of this super-rich soil that are six to eight feet deep. The yellow clay nearby is infertile and almost useless for farming – but terra preta soils support crops year after year. The evidence of terra preta being manmade is overwhelming, as there are bones and pottery shards throughout the soil, pointing to human intervention.

Yet how did the Amazonians make terra preta? Was it

created on purpose, or were they a side effect of long-term human habitation?

One could imagine pits being dug to dispose of waste, including leftovers, human excrement, broken pots and perhaps burned brush from clearing land. Or perhaps all wastes were burned in fires started on the surface of the ground and over centuries the soil built up beneath. Or perhaps terra preta was deliberately engineered, with bones, ashes, broken pottery, and charcoal all playing specific roles in soil fertility.

In last year's garden, I attempted to recreate terra preta by incorporating all the known ingredients together to see if I can also make soil that remains fertile for ages. To do so, we dug a pit and threw in fired clay shards, charcoal, bones, compost, and clay we dug from a creek. So far, my plants are responding well. The vegetables grown on the surrounding ground are not particularly happy unless regularly fed, but the terra preta garden bed and anywhere I have improved the soil with charcoal is doing better.

## Biochar

I have terrible sandy soil in my current garden, but I also have two smart friends named Steve – Steve Solomon and Steven Edholm, the former being the author of many gardening books and the latter being the creator of the Skillcult blog and YouTube channel. Both Steves recommended I add biochar to

my garden in order to improve it over time and to increase the exchange capacity of my sand.

Biochar is just charcoal made from various plant material. Hedge prunings, felled trees, even bamboo or coconut shells can be used. Last time I was in Ft. Lauderdale working on my Mom's gardens, I made a bunch of char in her backyard fire pit, mostly from coconuts, mango branches and bougainvillea prunings. Just be careful, though—some trees, like Brazilian pepper and oleander produce toxic smoke.

I did a good biochar burn in my mom's backyard fire pit, right in the middle of a Ft. Lauderdale neighborhood. I even burned green prunings from her bougainvillea hedge. My 5-year-old nephew Wyatt helped me, gathering sticks and coconut husks to put in the fire. We burned in middle of a weekday when it was unlikely to attract much attention. The fire pit is a decorative metal basin, suitable for a city backyard, and it still yielded about 15-20 gallons of char. You could do something similar in a backyard barbecue grill and no one would bat an eye. Heck, you could have some hot dogs on hand in case someone got too curious about the smoke. "I'm grilling over here, see?"

If you're really in a restricted area, you can skip the burning altogether and just buy hardwood charcoal from the store. Not the compressed briquettes but the real wood natural charcoal, still shaped like pieces of wood. That's will work just fine, though you'll have to smash it into smaller pieces.

To make your own biochar, just light your fire and get lots

of woody material burning well. Once it's starting to get white and has burned down to embers, put the fire out with the hose and harvest the black charcoal pieces.

## Charging Biochar

Once you have your biochar it's important to charge it by letting the charcoal soak up minerals so it doesn't eat up the nutrients in your soil. If you plow biochar right into a garden bed, it soaks up lots of fertility and renders your plants very unhappy for a year or more.

I had very good success soaking my biochar in Dyna-Gro, which is a balanced liquid fertilizer used in hydroponic growing. It has 16 elements in it which provide all that a plant needs to grow. Jack's or Miracle-Gro or any other soluble fertilizer should work too. If you are "iffy" on chemical fertilizers, you can use a lot of other amendments to fill those mineral gaps. I've added fish emulsion to get extra trace elements and some biological activity, along with a quart of kelp meal as well as a cup or so of pink Himalayan salt. All the minerals of the ocean right there. And speaking of the ocean, if you can get some seaweed and throw it in with your biochar in a bucket of water to rot, that should do it too.

You can charge your char with a variety of free things, including soaking it in manure tea or urine or even putting it in a bucket with alfalfa pellets and water to rot. You can also take the long approach and add char to your compost pile, letting it absorb lots of nutrition while also being colonized by bacteria and fungi. With my last batch of biochar I dumped the charcoal into a drum of Dave's Fetid Swamp Water, which is an anaerobic compost tea I make regularly. More on that in a minute.

Make sure you soak your biochar for a couple of weeks at least, then you can spread it on your garden beds and fork or till it in, or include it in a terra preta experiment.

If you plant fast-growing woody plants, you can get an endless supply of wood for making more char. Just crop them back and burn what you cut, then let them grow back for a year and do it again. Your hedges are great sources for woody material for charring. I turned a lot of bougainvillea prunings into char.

I would add about a 5-gallon bucket of charged biochar per 25 square feet of bed space for starters. Mix it into the soil to a depth of a foot or so. We are still experimenting with the exact amounts, but adding lots of charged charcoal makes everything happy in bad sand. I don't think you'll go wrong adding a little more or less.

## Adding Clay

I sometimes add clay to my sandy garden beds and my compost piles as a way to make the soil hold onto water and nutrients. Clay makes humus "stick." You can get clay easily by buying clumping unscented cat litter, which is just bentonite. Sprinkle it over your garden beds and see how it changes the tilth. You can also dig and bring some buckets of clay soil down with you when you visit family up north. I would love to have some North Carolina clay from Appalachia to add to my garden beds. Make a slurry and add clay to the sand along with charged biochar and you might actually get something resembling soil.

## Growing in Containers or Raised Beds

If digging pits and burning wood is not your cup of coconut water, you can always escape the sand completely by container gardening. Container gardens are excellent for herbs and vegetables as well as some small fruits.

You can make very nice container gardens or raised beds with cast concrete or stone. You can also buy barrels and pots or use old bathtubs. Horse troughs are great but expensive. Stacked cinderblocks work. Great big nursery pots will also serve you well and are much less expensive. Instead of getting the designer terra cotta or faux stone pots, you can get the big black plastic multi-gallon pots from a nursery supply company for a few bucks each. If you have a landscaper friend, you might even get them for free.

Raised beds can cause problems in Florida due to how fast the soil drains—especially if they're just rectangles of pressure-treated lumber on top of the native ground. They dry out fast—even faster than the surrounding soil! There are a couple of ways to improve this. First, you could dig deeper into the ground beneath the bed and throw in banana stems and/or chunks of wood to rot down and hold water. Alternately, you can go higher with your beds, making them perhaps 2' tall, then filling the bottom with wood and/or banana stems. Banana stems are really good water reservoirs.

Another method I've done that works well is to fill the bottom of a bed not only with chunks of wood, but other yard and kitchen waste, including food scraps, grass clippings, paper shreds, old non-synthetic clothing, hedge prunings, coffee grounds and filters, etc., and then putting about a 6" layer of soil on top of that. We do the same thing with large containers except we use potting soil instead of the native soil for the top 6".

Not only does this increase the water-holding capacity of the soil, it also increases the fertility of that garden. If you are buying soil, it saves you a lot of money because you only need a top layer. You can also throw a bunch of sand into the bottom part of the container/bed with your kitchen and yard wastes to fill in the gaps before you add good soil on top.

Potting soil can be bought in big scoops from some landscape and nursery supply companies and that is a cheaper way to go than buying bags of potting soil. I do not recommend buying bagged manure or mushroom compost, or municipal composts, or adding hay or straw to your beds, as all are potentially contaminated with garden-destroying herbicides. If you get those in one of your beds, you may kill your garden for a long time and basically have to start over with new soil or wait a few years to grow anything. The risks are not worth it and these herbicides are everywhere now. Potting soil is usually safe, though.

Five-gallon buckets can be pressed into service as planting pots, as can old kitty litter buckets. They're not pretty, though, and I would almost rather work in the sand than clutter up my yard with junk. It's better than not growing anything, but I believe a beautiful garden should be an attractive and peaceful escape as well as producing food. I'm done with tires and old fridges and toilet tanks and buckets.

I don't do much gardening in containers anymore, except in my plant nursery, but there is a place for it, especially in South Florida sand. It's nice to be able to make a small, fertile area and plant it in the same day. That said, you really *can* garden in just the sand.

## Just Pretend the Sand is a Hydroponic Medium

I was talking with Steve Solomon a few years ago about the unique gardening conditions in Florida. He told me that instead of thinking about the soil here as something worth building up, just consider it as something to hold plant roots in place. Fighting geology is hard, but you can grow crops just fine in the sand if they get enough nutrients from somewhere else.

"Think of the soil as a hydroponic medium," Steve said. "Instead of feeding the soil, foliar feed the plants."

This is actually really easy. Just make your beds and plant them, and if you have some compost or fertilizer at the beginning, feel free to add it—but from then on, most of your feeding will be via soaking the plants with nutrients during watering. An easy way to do this is to make Dave's Fetid Swamp Water (DFSW) from various plants, or feed with a purchased water-soluble fertilizer like Jack's Plant Food, or, if you're a weirdo, use diluted urine (6-10 parts water to one part urine), or mix up some fish emulsion, or make teas from moringa, poinciana, *Tithonia diversifolia* or other leaves.

You can learn in depth about DFSW in my book *Compost Everything*, but I'll relent and have pity on you, oh dear reader, so you don't have to go buy another book, even though it's amazing and transformative and is the best book on composting that you'll ever read in your entire life even if you lived to be two hundred and fifty years of age.

"Dave's Fetid Swamp Water" is a free liquid fertilizer you make by throwing a mixture of materials into a bucket or barrel and covering them with water to rot down for a month or so. You can rip the weedy vines off your fence, put them in water to rot, then water your garden with the stew they make after rotting down. Or you can make the mix better by adding leaves from various trees—especially the leaves of nitrogen-fixing species like necklace pod, cassias, poinciana, etc.. Even better, add some fish guts and shrimp and crab parts. Better still, throw in a few cups of Epsom salts for magnesium and sulfur. And some seaweed. The mix will rot together and smell terrible but it does a great job growing plants. I recommend applying it when the neighbors are NOT outside. The smell will go away in a few hours, but it really reeks. That's how you know it's good! It ain't called "fetid swamp water" for nothing.

It also doesn't spoil. You can make a batch and use it in two weeks—or two years!

But, DFSW aside, the idea with liquid feeding is that you're soaking your crops and the sand with a nutrient solution, "fertigating" your garden and giving it what it needs. It's way easier than trying to improve the sand and making tons of compost to try and fight Florida's geology. Between the heat, the rainfall and the sand, it's super hard to keep life in the soil or maintain fertility. So if you don't want to fight it, stop trying to fight it. Just plant crops and water them regularly with a fertilizer solution.

## Watering

South Florida gardens need a lot of water. Most of our rainfall arrives in the summer with some long, dry patches in between. The number one thing plants need is water. If they aren't getting it regularly, they suffer and produce poorly. Sometimes plants yield a harvest without great nutrition, but they won't yield much of anything without much water. Think about how fast you get thirsty working outside in the Florida sun. We need water more than we need food and can go for much longer without eating than without drinking. Your garden is the same way. Regular watering is the key to a happy garden.

And when you water, water deeply. It's better to water deeply a few times a week than it is to give everything a shallow watering daily. Shallow waterings may moisten the top of the soil but they don't go down very far. This not only keeps the plants from getting what they need, it also teaches them to keep their roots near the surface where the water is, making them more susceptible to drought. If you water deeply, as the soil dries out in between waterings the plant roots will grow downwards and chase the water you've already given them.

Sprinklers are my favorite way to water in Florida. Drip irrigation and hoses fall apart quickly, clog up and get gnawed by rodents. A few sprinkler heads on stands work great. Watering in the early morning is probably best, though you can really water any time you like. Some plants—I'm looking at you, tomatoes—can get fungal issues if they're watered in the evening and stay wet all night. That said, it's better to water thirsty plants at the "wrong" time than it is to wait until another day. I've heard people say that you shouldn't water in the middle of the day because the water gets heated up by the sand and will cook the roots but I've never found that to be the case. In fact, I think they like being cooled off a bit and I really doubt the water stays warm all that long. People say all kinds of weird stuff.

Small gardens are easy to water by hand with a garden hose, though remembering to do it can be a problem. If you're really not good at keeping your gardens watered, consider putting up some sprinklers and a timer.

Incidentally, if you want to learn how to garden without irrigation and use the water you have available in the most efficient manner, I highly recommend getting a copy of Steve Solomon's recent book *Water-Wise Gardening*.

## Plant Northern Favorites in the Right Season

Though some of our northern favorites like carrots, tomatoes, cabbages, and peas aren't ideally suited to South Florida's climate, they can be grown in your garden. The trick is to grow them when the weather is cool. Even traditional summer favorites like sweet corn, beans and tomatoes like cooler weather and less humidity than Florida provides during the hot months.

For example, the University of Florida's *Planting Guide for*

*South Florida Vegetables* notes that South Florida gardeners can plant the following crops in *December*:

> *Beets, Broccoli, Cabbage (regular & Chinese), Carrots, Cauliflower, Collards, Corn, Cucumber, Eggplant, Endive, English & Southern Peas, Escarole, Kohlrabi, Lettuce, Lima, Pole & Bush Beans, Mustard, Onions, Parsley, Peppers, Potatoes, Radish, Spinach, Strawberries, Tomatoes (larger fruit varieties), Turnips*

Yes, that's right. You can plant all those popular warm and cool-season vegetables before Christmas. That's not the way my great-grandpa gardened in Upstate New York! Trust me, if you try to plant your vegetables "in the right season" as written on seed packets, the Florida summer will punish your unwitting ignorance.

For the complete recommended planting times for various vegetables, I highly recommend following the South Florida portion of the University of Florida's "Florida Vegetable Gardening Guide" which is available online as a free pdf. Print it out and stick it on your fridge!

There are some traditional vegetable crops that will live in a Florida summer, like Southern peas, okra, and sweet potatoes, but there aren't many. Your main gardening season is fall, winter and spring. Summer is mostly a dead zone for "traditional" gardening, which is fine because you won't want to be out in the heat anyhow!

## Where to Plant a Garden

Most vegetables prefer "full sun." Up north, this means sunlight throughout the day, as in, plant your vegetables in the open where the weak and watery light of the arboreal climes may coax a harvest from them. Down in South Florida, full sun is not needed for most crops. Half is good enough. If possible, try to place your garden so it gets most of its sunshine in the morning rather than the afternoon. If you get a good four to six hours of full sun in a day, you'll be fine. That said, it's best not to plant gardens right under trees where your vegetables will have to fight with tree roots to get the water and nutrition they need. Also, don't plant your garden in a wet or swampy area, unless you're growing taro, water chestnuts or rice.

You should also plant a garden where you'll look at it. Don't build your garden in the back of your yard by the old rusty shopping cart and the broken kayak. Instead, put your garden where you live. If you sit on your back porch in the evenings after work, plant the garden right off the porch where you can look down and see it. If you commute to work daily, plant your garden by the driveway so you walk past it. My sister Linda recommends planting a garden you can see from the kitchen window over your sink "so you can be reminded daily that you are behind on dishes AND gardening. It's fantastic." If you see your garden during your daily routine, you will head off potential problems. You'll see plants wilting. You'll notice insect issues. You'll discover the anaconda in your carrots. The garden becomes a part of your life. If you instead plant in the back corner of your yard, you're less likely to spend as much time in the garden or catch little problems before they become good ones.

To recap: plant your garden in full to half sun, away from tree roots, in a non-flooded spot that you'll visit regularly.

## Making Compost

South Florida composting is easier than it is up north as we have a warm climate that is perfect for the proliferation of composting organisms. You can compost year-round without worrying about a pile freezing or having to wait until spring to get things really heated up. My favorite method of backyard composting is to make a good-sized cube-shaped bin (ideally about 4' x 4' x 4') or a pair or even a triplet of bins, each with that triple-4 dimension. In the city it's a good idea to have a top on the bin as well to keep out possums and rats and whatever invasive reptile just came in on the last banana boat. I've made bins from pressure-treated lumber and from concrete blocks. I prefer the latter as it never breaks down. When you read books on composting it sounds like the process takes a post-doctoral degree, but it doesn't really. Just throw in organic material as you have it and water the pile well. Shredded paper, grass clippings, leaves, kitchen scraps of all kinds (including meat, dairy and bones), rotten fruit, seaweed, garden waste— just throw it all in! The wider the variety the better the future garden nutrition. My mom has a compost pile in her backyard that she feeds as she has material. She doesn't bother watering

or turning it. It's just a wooden bin stuck to the fence behind the shed—super easy! When I visit and tend the food forest, I push back the recent material on the top and dig out the rich, dark, earthy compost lower down in the pile. Florida practically composts for you, it's so easy. Don't get stopped by all the rules you see online if you're worried about how to compost. Just throw organic material in a bin and let it rot. I like to run the mower over fallen leaves and bag them to add along with grass clippings and fallen mangoes. Water that mess well and it will compost down quickly. I have a rectangular frame of 2 x 4's that has hardware cloth on it I can use for a sifter when I want fine compost for seedlings and new beds. We put it over the wheelbarrow and sift away—easy as pie. For even better and more long-lasting compost, I recommend throwing some clay into your bin. Just chuck in handfuls of clumping unscented kitty litter or whatever other natural clay you can find every time you add some material to the bin.

## How To Make a Garden Bed in 30 Minutes or Less

To easily create your first garden, mark out the spot where you want to plant. I prefer to make beds that are 4' wide and as long as I can conveniently build them, with 2' wide paths in between. In this year's garden, however, I have a set of beds with a mixture of trees, berries, and vegetables which I placed 3 feet apart. (I call this my Grocery Row Garden system and it is an ongoing permaculture experiment that involves overlapping an orchard and a vegetable garden—it's a lot of fun).

In most South Florida backyards those 3' paths might be too much "wasted" space, though. With smaller vegetables like carrots, bush beans, green onions, and radishes, you can get away with cutting your paths between beds to about 18", though that is the absolute minimum width I would give my paths.

Using stakes and twine, mark out the boundaries of your first bed and remove the grass. A shovel works fine for this as South Florida sand is easy to dig, unless you have limerock. In that case, you either need to use a mattock to chop holes in the ground and add soil or go up by building raised beds as previously discussed. I recommend having at least 12" of soil depth.

You don't need boundaries on your garden beds (limerock excepted), as raised beds with wooden boundaries are usually not the best way to garden in Florida sand. Remember, raised beds require more watering and get hotter than in-ground beds. They're also harder to weed around as the borders get in the way. With the simple beds here, you just work directly in the ground without spending money on lumber that will rot out in a few years.

Mark out the space you want to plant by using twine and stakes or just rocks at the corners, then take out the grass. Removing the grass is usually just a matter of chopping at it with a shovel and pulling it up in clumps, shaking the sand from the roots as you go. Set aside the grass (or weeds) you pull out so you can compost them. I recommend first throwing them on a concrete slab or your driveway to dry out for a week or so so they don't re-root in the compost.

Loosen up the sand with a fork or a broadfork if you have it and throw in whatever rich organic matter you can find. Leaf mold, coffee grounds, compost and alfalfa meal are all good choices. A couple cups of 10-0-10 can help too, but you'll have to keep adding it again and again because it just washes out. I don't recommend buying compost or manure—or getting it for free—as many of them are contaminated with long-term herbicides (look up "Aminopyralid") used to control weeds—they'll often kill your garden before it even gets a good start. Multiple Florida gardeners have told me they lost entire years because of spreading around manure or rotten hay. As I wrote in *Compost Everything: The Good Guide to Extreme Composting*:

> *Don't import manure for your garden if:*
>
> *The source farm isn't organic*
>
> *The animals are eating imported feed/hay or living in imported bedding straw The animals are treated with chemical de-wormers/antibiotics/etc.*
>
> *A bagged manure/compost contains "biosolids"*
>
> *As you'll quickly see from that list it's basically impossible to meet those criteria. Yet if you don't, you're running the risk of poisoning your ground. It's just not safe to add manure anymore unless you know it's*

*safe…. Don't bring manure, compost, straw, or grass clippings onto your property. Trust no one except people that don't feed their animals any purchased hay and who you are sure do not spray their fields with anything. This is the only way to be completely sure your garden won't get whacked. Look, I'm not hyper cautious, but this is deadly stuff, and it sticks around in the ground. It's been years since I got hit and many of my perennials never recovered. The supply chains are really long.*

*It's really hard to find out where hay and straw originally came from. Chances are, a lot of it is being sprayed.*

*Aminopyralids don't hurt grasses, so they're often used on wheat, corn, grains, and pastures. In the name of convenience and saving time, they're poisoning the supply chain for organic farmers. Once you know about the existence of these long-term pesticides and the range of their use, you'll look sideways at a lot of amendments that used to be perfect for your garden. The game has changed. Don't get nailed.*

Make your beds right in the ground the easy way and don't worry about the fancy stuff. And don't kill them with manure. A half-inch of compost forked into the ground is enough to grow vegetables for one season, though you will see greater yields if you also add foliar feeding. If you don't have compost, turn under a bucket of alfalfa pellets if you have them. Rake the top surface of your beds even and round off the edges and—voila—you have a beautiful little mounded bed of sand, ready to plant.

Now that you have your garden beds, let's plant.

CHAPTER 4:

# Balancing the Tropical and the Temperate

New South Florida gardeners often rush out to buy seeds for the plants they're used to growing up in Yankeeland. Things like sweet peas, beefsteak tomatoes, carrots, and beets. They also long for apples, pears, peaches, and other crops and wish they could grow them.

But Florida laughs at these crops and smites them in her pettiness. That isn't to say that all of these fruits and vegetables won't grow for you—just that they won't grow like they do up north. If you try to plant a traditional Victory Garden in Miami, you have to really work at it. I spent years growing normal garden vegetables in South Florida before I got smart and started growing what really *wants* to grow in this climate, rather than what I *wish* would grow in this climate.

If you base your gardening on vegetables that are suited

perfectly to South Florida, you will have success right from the beginning, even if you aren't an expert gardener.

## Grow the Easy Stuff First

These are just a few plants that will get you started.

If you grow bananas, plantains, cassava, yams, sweet potatoes, and dasheen for your starchy calorie crops, you're gardening on the easy setting. Add in some chaya, longevity spinach and edible-leaf hibiscus for greens, plus some snake beans, okra, Seminole pumpkins and ivy gourds and you'll have lots of great food for the table with very little work. At the edges of your yard, plant bananas, plantains, various small fruits, some bigger tropical fruit trees and jicama and chayote squash. Some of these plants grow so easily they are considered weeds by the state of Florida.

Don't be afraid to try new things! With some patience and trial and error, you'll discover that South Florida is a great place to grow enjoyable and delicious food, even if it isn't what you are familiar with. We can tap into the fruits and vegetables of the entire tropics.

I will cover my 25 favorite South Florida food crops in depth in Chapter 6. If you base your gardens on these suggestions, you will have plenty of food.

## Planting

First, consider how you want to water. If you're not going to set up automatic sprinklers, you'll be watering by hand. If

you mulch your garden, you'll have to water less often than if you leave it bare sand. Also, if you space plants farther apart you'll need less water. Tight plantings take more water, widely spaced plantings take less. Plantings in full sun need more water than those in part shade. I usually just draw some furrows in my bed with the corner of a hoe and drop seeds in, then lightly cover them and water well. Larger plants that need more room get planted in individual holes with generous spacing. In South Florida you don't need to make mounds for planting root crops like cassava and yams—you can just stick them into level ground and they'll grow just fine, unless you have unusually compacted soil. In that case, loosen the ground first.

If you want temperate climate crops to thrive, you need to grow them in late fall and winter. Peas, carrots, beans, tomatoes, bell peppers, sweet corn—all of these can do well in a South Florida winter, though they're a little needier than the tropicals I mentioned above. Our winters are like spring up north, so think of what would thrive during May in New England and you know what to plant during South Florida's coldest months. Unfortunately, the winter is also Florida's dry season, so you'll have to keep everything watered. Having a rain barrel is nice, as rainwater is better for plants than chlorinated tap water. Bonus points if you have a well.

Once you hit May-June, most of your normal temperate vegetable gardening should be done. At that point, however, your tropical plants really start kicking, especially as the rains start in June. Cassava, yams, sweet potatoes, dasheen, longevity spinach, chaya, snake beans, okra and other warm-climate crops will explode into growth with the rainy season, giving you an entirely different range of produce for your table. You won't have any luck growing lettuce or cabbage or carrots in summer, so wait until late fall or early winter to plant them again. Hot peppers are a year-round perennial in South Florida which you can plant any time. A cayenne pepper bush can get as tall as you if it's happy! Unfortunately, bell peppers don't do nearly so well.

Learn to plant and grow the crops that like Florida and make them your primary vegetables through the year, then work on your "normal" temperate species during the cooler months.

## A Final Note on Feeding Your Garden

As I mentioned above, you do need to add something to the garden to feed your plants. Get those liquid fertilizers going. You can use 10-0-10 or other granular fertilizers, but they leach through fast. I would also add micronutrients if you're using chemical fertilizers. Epsom salts are great when sprinkled across the garden. Seaweed from the ocean makes a good mulch that slow-releases nutrients. I usually rinse it first, except when I put it around trees or tomatoes. In that case, the salt doesn't seem to hurt anything.

Stay on top of feeding and watering and your plants will grow fast. If they look wilted in the morning or are getting yellow, you aren't feeding and watering them enough. Treat them like babies. You wouldn't do well in the Florida heat without food and water and they won't either.

But enough about vegetable gardening for now. There is an even easier way to grow food in Florida.

CHAPTER 5:

# Grow Beyond the Bed (Fruit Trees Are Your Friends)

I am convinced that perennial vegetables, trees, and herbs are the very best choices for a lazy South Florida gardener. It takes work to grow tomatoes and lettuce, but it takes almost zero effort to grow mangoes and jackfruit, ginger, and cassava. Carrots and peas take effort. Tamarind and moringa don't.

If you think a little longer-term, you can really grow with South Florida's climate by transforming your landscaping into an edible paradise.

If you haven't thought about adding tropical fruit to your gardening, you are missing out! Have you ever had a good starfruit? Some people don't like this charming fruit because they've tried a sour one, or worse, a bland and watery one with bitter skin. A great starfruit is a perfect lemonade-esque balance of tart and sweet, juicy and refreshing on a warm day. The tree is also very attractive and quite easy to grow.

Let's start with a starfruit. Grab a tree and put it in the ground, then plant some other edibles around it. How about a guava? And a dwarf banana? And some cassava? Throw in some perennial African blue basil to bring in the butterflies, and maybe some pentas or lantanas for color and attracting more pollinators. Water well, then put down cardboard over the remaining grass between the plants and pile up a foot or so of mulch. Throw around some slow-release fertilizer and keep the area watered for a few months until it gets established. Now you have a lovely little island of edibles that will feed you for years with very little work except for occasional pruning and picking and mulching.

Do you have a gutter running into your yard? Plant bananas or plantains where it drains and they'll love that extra water and give you fruit.

Need a hedge? Plant guavas or cocoplums or Surinam cherries. If you can find black Surinam cherries, they are a really good fruit. Cranberry hibiscus (*Hibiscus acetosella*) is another attractive hedge plant with tart, edible leaves.

At the base of your hedge, plant pineapple tops. They take very little work and once they get established, they'll make pineapples for years. Those pineapples will be way better than store bought, too. Fresh, fully ripe pineapples are a treat.

Do you want to grow your own multivitamin? Plant a moringa tree! They grow easily from seed and can hit over ten feet in height in a single year. The leaves are very nutritious and can be harvested to make tea or to be stripped from the branches and added into soups and sauces for a dose of extra vitamins and minerals.

Do you have a trellis to cover? Plant passionfruit or perennial cucumbers, true yams or jicama—or all the above. These perennials will happily create shade and privacy while yielding food for your table.

For a lovely specimen tree, plant a jabuticaba instead of a crepe myrtle. For a lovely shade tree that bears amazing fruit, plant a star apple instead of a gumbo limbo.

I don't understand why South Floridians keep planting the same silly non-edible trees everywhere. Look, I like gumbo limbos as much as anyone else, but come on—we don't need any more of them planted in our limited real estate! They get popped in everywhere, just like oaks and poinciana trees. Yeah, they're pretty and all, but plant a mango or an ackee or a mulberry for goodness' sake. We've got an awesome climate and can reach deep into the tropics and grow stuff the rest of the nation could only dream of growing ... and you plant a gumbo limbo? Plus, our real estate costs bazillions of dollars. The square footage of your yard should at least try to pay for itself.

Plant.

Fruit.

Trees.

And the next hurricane that comes through, chop down

your ornamental trees and compost them so you can plant lychees and June plums instead.

Whew. I get worked up about this. South Florida has fantastic growing potential. Don't waste it. If you plant the average South Florida yard from stem to stern with fruit and nut trees and perennial vegetables, with some annual beds thrown in, you could feed the whole block. Maybe one day we'll have gardeners on every street using the incredible climate to its full potential. That's my dream. As I wrote in the introduction to the first edition of *Create Your Own Florida Food Forest*:

> *Imagine transforming your yard into a Garden of Eden. Fruit trees sway overhead, berries and flowers burst forth from the shade, and alongside soft paths sweet potato vines intertwine with passionfruit and native wildflowers. Though the air, bees and butterflies buzz and flutter, spreading pollen and beauty in their cheerful wake.*

> *You can do this in less time than you think. You don't need to over-plan or over-think. Spacing isn't all that important. You can start with seeds, cuttings or potted trees. You could even plant a decent food forest just by visiting your local international market and collecting seeds and roots to plant.*

> *Your limitation is your imagination. Florida WANTS to grow forests. Before development and clearing, our*

> *region was a verdant jungle—a land of flowers. We can create that flowering jungle again—and tailor it to serve us by growing plants and trees that provide food, beauty, wildlife habitat, building materials and fuel.*

Florida should be a paradise. Grass and oleanders may look nice, but Eden was filled with fruit trees. Eden is better.

If you want to see what South Florida could be growing, go visit the Fruit and Spice Park in Homestead—then go hit a fruit tree nursery! Take care of those trees for a few years and they'll feed you for a lifetime.

Though it's worth vegetable gardening in South Florida, the region really shines as a home for trees and perennials.

We're so close to the tropics you can literally taste it. Pineapples, mangos, coconuts, allspice—the possibilities for culinary delight are endless. You can grow the plants that the rest of the nation can only dream about.

Unlike vegetables, tropical fruit trees are quite easy to grow in South Florida. I planted a food forest in my parents' yard a decade ago and despite infrequent watering and fertilization, it's producing an abundance of fruit. With a little push, it could feed multiple neighbors as well as my family. Because vegetables don't need full sun to thrive in South Florida, trees and veggies don't have to be mutually exclusive. Your orchard can also be your vegetable garden. When we lived in the Caribbean, we saw banana and plantain trees surrounded by cassava, sweet potatoes, cabbages, and tomatoes, while coffee, cacao, mangoes, cinnamon, and other tropical fruit trees thrived at the edges of the field. If you're tired of all the work that goes into growing vegetables, plant fruit instead. Plant jackfruit, mangoes, plantains, ackee, mulberries, and more. They'll grow into productive trees that take very little work.

An acerola cherry tree will give you sweet-tart fruit through much of the year, also providing you with plenty of vitamin C. Grow some chaya and you'll have all the cooked greens you can eat. Plant pineapple tops around a coconut palm and you'll have piña coladas for life.

Fruit trees can be daunting to the beginning gardener. He looks at a display of trees, thinks about the size of his yard, wonders about how long it's going to take, gets afraid the tree will get too big or will take too long, then goes home without the tree.

Don't worry, oh timid gardener! Just don't plant a tree right next to your house – and remember: you CAN control the size of a fruit tree. Just because Mr. Google says your tree can grow from "35 – 100'" tall doesn't mean you need to let it get that big.

Another tip on South Florida tree gardening: those maximum sizes almost never happen in our sandy soil and with our somewhat cooler climate. A tree that might grow 50' in Brazil usually tops out at more like 25' in South Florida. The tree might still get too big for your yard if you don't prune it, but it's unlikely to reach anywhere near the top of its genetic potential.

It's easy to keep a tree from hitting its full growth potential. All you need to do is cut it back once or twice a year. Buy a pruning saw and cut, cut, cut. If you want your mango to stay at 8' tall so you can harvest mangoes without a ladder, then keep cutting it back until it gets used to being short. You won't kill it.

Have you ever seen a ficus hedge (*Ficus benjamina*)? People often hate those hedges because when untended, they can get huge. Yet when pruned regularly, a ficus hedge is a nice, dense block of green that keeps the neighbors from seeing you taking selfies on World Naked Gardening Day. However, if you take one ficus and plant it in the middle of your yard and just let it grow, it'll turn into a massive tree with dangling roots hanging from the branches. They're a great climbing tree for kids—we used to climb them all the time in the field behind Steven Foster Elementary School—but they are huge. Yet when planted tightly together in a hedge and pruned, they're totally controllable. I was visiting a friend on Sunday and saw a ficus tree in a pot in his sunroom. It was thin, a bit leggy and green, growing in maybe a 30-gallon pot. It wasn't 60 feet tall and putting roots down through his chimney while smashing thick branches out his plate glass windows. The tree will never reach its genetic potential. It has been constrained by its owners.

You own your trees. Make them work for you. Prune them at the peak of their growth during South Florida's rainy season and you'll lower their vigor. Use the prunings as mulch to feed your other plants. Keep those trees under control! It doesn't take much work and it doesn't take a genius. It just takes a pruning saw and/or a pair of loppers. You can fit a lot of trees in your backyard if you're willing to train them – much more than you think. Fruit trees can be made into hedges, planted in swales, grown in large pots on patios and more. Lose your fear. Teach them to grow how you want them to grow.

My in-laws have an old mango orchard in their yard that was probably planted back in the 70's. Those trees are as large as oaks, dropping mangoes from a terrifying height. In their shade the grass is thin. When my wife was a teenage girl, she had to mow the lawn beneath those trees. In mango season, she got splattered with rotten mango pulp, giving her an aversion to mangoes that was only cured when she moved to the Caribbean and started trying new varieties from the local markets.

When you have a giant tree, it may look beautiful, but it's not necessarily the most useful to you as a food grower. Fruit falling from high up smashes on the ground. Yes, you may get a literal ton of mangoes – but many of them are destroyed. It's often better to grow smaller trees that you can easily tend and pick. If the fruit falls from eight feet instead of sixty, it's much more likely to survive the impact with the ground. It's also easier to protect the tree from pests and disease. When the canopy requires a helicopter to inspect, there really isn't much you can do.

Here's another thought: that giant tree is filling up a space in which you could plant six smaller trees. Instead of one variety of mango, you could be growing six. Or you could grow a mango, a starfruit, an acerola cherry, a mulberry, a black sapote and a jaboticaba! Think about the variety! The extended harvest seasons! The bragging rights!

When you plant trees in South Florida sand, water them in well to get rid of air pockets. Then make a little dam/basin around the base of the tree that you can fill with water a few times a week. It's easy to do this in the sand.

Keeping a tree small should start when you plant it. You can lop half the tree off at planting if you like. It won't die, provided you keep it watered. Lack of water is the #1 tree killer in South Florida. They need to be cared for in our sand or else they give up.

A backyard full of trees makes for an easy garden. Just prune and pick, water as needed and enjoy the bounty of the tropical rainforest that South Florida is supposed to be.

Now—that said—there is a class of trees that don't always do well in Florida anymore. Citrus. It's a terrible thing, but oranges and grapefruit and other citrus are no longer easy Florida gardening trees.

A few years ago, I went orange picking at a Florida U-Pick grove closer to the middle of the state. At first glance, the scene was idyllic.

A Victorian-era home with a friendly wraparound porch and an outdoor barn sat near the entrance to the grove.

Five-gallon buckets of citrus sat on the ground for sale and the elderly proprietors, a man and his wife in their 80s, waved as we pulled up.

"The tangerines are mostly gone and the grapefruit aren't in yet," the wife said as we stepped up to the table with the cash box. "You can pick all you like of the oranges, though."

"What types do you have?" I asked, curious.

"All different kinds. I can't even tell you anymore," she replied. "Both juice and eating oranges. All good."

I thanked her and set out with my son through the grove. Above were a few stately pecans, overshadowing both thorny seedling trees and well-tended oranges.

There were all sorts of oranges and every single one we picked turned out to be delicious; yet as I wandered the grove, I saw quite a few trees with yellow leaves and less-than-healthy growth. A few were half-dead and some spots had recently been filled with new trees. It was a beautiful grove at a distance—yet up close, all was not well.

As I walked around, I decided to film the fruit, the trees, and the beauty of the grove. I thought to myself: will my children see a grove like this ten years from now? Or in twenty years?

We filled three buckets (the cost per bucket was only $6 so why not?) and walked back to the table in front.

As I checked out, I asked the woman "Have you been having problems with citrus greening?"

She nodded. "We planted this grove a long time ago. Now

I don't know if it's going to be around in even a few years. Lots of the trees got it. It's not good."

I shook my head, offered my condolences, paid with a $20 and told her to keep the change.

It hurt to see those trees and that couple under the cloud of an incurable disease.

Few things represent my home state of Florida more than oranges. They're a symbol like few others can be. They're definitely better loved than alligators.

Yet thanks to citrus greening, the orange industry is falling to pieces.

The spread of citrus greening means the tree you buy and plant today is likely to end up dead within a decade unless something changes quickly.

It kills me to say this, since I love my citrus trees and wish I could plant a dozen more—yet the psyllid that has infected the groves is known to travel for miles. That means if you're in or near a greening infected zone, you're likely to end up with the virus before too long.

One of the more painful things I had to do over the last decade was to pronounce the last rites over my Mom's young Navel orange tree. She was so happy to have that tree given to her, but only a few years after planting it went into complete decline, bearing twisted fruits and yellow leaves.

I've heard similar stories of citrus trees that were planted in greening zones and rapidly succumbed to the virus.

It's all across the state and if it's not in your area, it's likely to spread there.

To say that I'm really sad over the spread of citrus greening would be an understatement.

I want this thing cured. I love citrus trees. There is some promising research on greening, but I have no solid recommendations on beating it yet.

When I was a kid, we had a huge grapefruit tree in our backyard and my Dad built a tree fort for my brother and me in its branches. There were so many grapefruit we could hardly give them all away.

That's the Florida I want to see again. I fear it may not return, yet I still hold out some hope.

Right now, though, I'd skip planting citrus. There are many other trees that make more sense and don't have a viral sword hanging over them.

Another tree that people complain doesn't fruit well enough is the banana. Unlike citrus, the problem isn't a disease. The lack of abundant fruit and happy trees is a cultural problem. Bananas like fertile soil and lots of water. To get bunches of bananas and plantains is actually easy. All you need to do is water regularly and feed them and they'll produce like mad.

Remember, South Florida is a tropical climate. As Bill Mollison writes in his *Introduction to Permaculture*:

> *In the tropics, it is possible to be food self-sufficient*
> *from trees within two or three years. You start with*

*things like bananas and papaya, and go on to a huge variety of fruits and nuts. There are lots of staples, too, like a coconut. Back about the 1940's, the coconut was fully used. "The Pacific Islands Year Book" gives 467 by-products around a tree like that. Breadfruit produces so much food that it becomes incredibly wasteful! The breadfruit is quick to propagate, and easy to grow. I will tell you a little story. There is a man named Cliff Adam, living in a group of islands with about 40,000 people. Cliff got a grant from the United Nations to collect some food plants that might suit the area. They gave him $136,000. So he took off in his plane and kept sending home parcels. He left two or three friends there who kept planting all these trees. He sent back some 600 sorts of mango, 30 or 40 sorts of breadfruit, all sorts of guava, and so on. When he got back home, he then moved them out in rows on 68 acres near the shoreline. Then he got another 135 acres from the government, up on the hills. So he set out all these trees. About three or four years later, he had all sorts of cassava and all sorts of yams and taros that you could imagine. He said to me, "I am in a very embarrassing position."*

*I said, "What is wrong?".*

*He said, "Well I shipped this crop in that wasn't growing here traditionally." This was really a coconut economy. He shipped all these plants in, and he set them out as trials. So he said, "The problem is, what I was going to do was this: give the farmers different sorts of mangos, breadfruit trees, and all that, and I have been doing it; but already the production from my two hundred acres would feed the island, and that's experimental production. I am in the embarrassing position where, as agricultural research and nutrition officer, I am already alone responsible."*

*He said to me, "What am I going to do?"*

*I said, "I dunno."*

*This is a difficulty wherever people undertake this sort of assembly. You haven't gotten very far along the road, maybe four to seven years along the road, when you've grown so much food the whole thing gets rather embarrassing, and if you are the agricultural officer of a small country, you could probably feed the country on the experimental plots.*

The tropics are an amazing place and South Florida is functionally tropical. If you plant your little backyard with highly productive tropical species and you take care of them, you can just about feed your neighborhood in a few years. What a great problem to have, right?

You can start by planting one tree. Around it, plant some smaller edibles. Then plant a ground cover of sweet potatoes.

It will grow and thrive. When you see the beauty of that little island, you'll want to plant more—and more—and more!

South Florida isn't a gardening wasteland. It's an amazing opportunity if you approach it with new eyes.

## A Final Note

Though you are technically allowed to grow food anywhere, thanks to Florida Statute 604.71, you may still have to fight for your right to cultivate. Fortunately, you can hide a lot of food in plain sight. A hedge can be grown from guavas or Surinam cherries or cocoplums or any number of other edibles, a clump of sugarcane is very ornamental beside the front door, a chocolate pudding fruit or a star apple looks lovely as a shade tree, Peruvian apple cactus tastes like dragonfruit but its columnar form is stunning in a modern pool-side design, and—you get the idea! A garden of edible food doesn't have to look like rows of vegetables. You have lots of options. Your yard can be both beautiful and productive.

CHAPTER 6:

# The Good Gardener's Top 25 Easy-to-Grow South Florida Crops

The more you read about gardening, the more complicated it seems. You must consider feeding and watering and timing and pests and iguanas. It's hard when you're just learning, but it's even harder when you start with touchy crops. Some plants are a pain, whereas others grow so well they make you feel like a gardening genius in your very first year of tending the soil.

Those plants are the ones we're going to cover in this chapter. If you plant these delicious varieties and pay even a little attention to them, you will have success. In this chapter I'm giving you the superstar plants of Florida gardening. Many of them may be new to you, but that's okay. Soon you'll know them by heart and will be enjoying sharing the taste of the tropics with your friends and family.

I'll give you a few growing notes on each plant, along with an idea of how to eat them. Ready? Let's grow!

## Acerola Cherry

Also known as Barbados cherry, this vitamin-C rich sweet-tart fruit is a delicious addition to your yard. The trees are scrappy and easy-to-grow. Some varieties bear off and on almost year-round and others have a main season crop. The fruit looks similar to a true cherry but has three winged seeds in the center instead of a pit. They are delicious right off the tree and are a favorite with children. My kids can't get enough of them. Plant your acerola tree in full sun and keep it watered a few times a week for the first few months until it seems established. Mulch around the base and throw down some compost as you have it. The natural inclination of the tree is to become a big ball of branches, perhaps 12-16' tall. Pruning won't kill it, so prune as needed. The best way I've found to eat this fruit is right off the tree. When picked, they only keep for a day or so on the counter.

## Avocados

Hass avocados are not generally recommended for Florida but there are many other good-tasting varieties. I recommend planting a grafted tree so you can get fruit fast. Avocados do grow readily from pits, but the time to fruiting can be excessive (up to a decade). Ask your local nurseryman about his favorite varieties. I like all the named varieties I've tried. Avocados ripen after picking and it can be a little tricky for new growers to figure out when they are ready. We watch and see when they appear full-size, then pick a few and leave them on the counter. In a week or so, they soften up (you can squeeze them and they give a bit) and are perfect to eat. Open them early and they are awful; open them too late and they are off-flavored and mushy. There is a day or two where they are perfect. If you pick an avocado off the tree too early it never ripens up properly. You'll learn by experience. Some types turn a little different color when ready, but that also varies by cultivar. We've seen some that are ripe at dark green, others that are a lighter green, others are purple or even reddish when ripe. Just pick a test avocado here and there and let them sit on the counter until you get ones that are ripening fine and taste good. The picking season lasts for weeks, so don't pick them all at once. Avocados are wonderful cut in half so you can remove the pit—just add a pinch of salt and scoop out the buttery flesh. Or make guacamole.

## Bananas/Plantains

All sorts of bananas and plantains grow easily in South Florida, with one important caveat: they need food and water to bear well! Bananas and plantains like about double the rainfall that South Florida provides. If you just stick a banana pup in the middle of your sandy backyard, it grows slowly and fruits poorly. If you put it next to a leaking septic tank, it will grow like a rocket and bear immense crops. (Ask me how I know.) Greywater drains, though of questionable legality, are excellent places to plant bananas. In our climate you could live on bananas and plantains. Ripe, they are sweet and delicious. Bananas are good off the stalk and ripe plantains are good fried, baked or roasted in a fire. If you pick bananas or plantains green, they can be peeled and boiled, added to stews, boiled and mashed into a porridge with sugar and spices, or sliced and fried in oil. It's like the best of a root crop and the best of a fruit crop—and perennial! Some banana trees mature quite rapidly and produce fruit within a year of planting if well-fed and watered. Every South Florida yard should have a clump of bananas. If you are along a canal, plant bananas near the water and their roots will pull up what they need. My wife's grandmother had a house on a canal in Ft. Lauderdale. Out back was a huge stand of Orinoco plantains that got watered daily when the tide came in. They thrived in that mucky canal water and there were always stalks of bananas growing, though Rachel's grandmother said they were "horse bananas" and weren't worth eating. I decided to test this hypothesis and harvested some to eat. Raw and ripe, they weren't great— but fried, they were delicious. Orinocos are pretty common in South Florida, though they aren't nearly as good as "normal" plantains. They are scrappy, though, which is a benefit, so I always plant some in my gardens. Bananas and

plantains can take some flooding so long as they aren't perpetually submerged in water. If you have a hot, dry area they will do poorly. If you have an area with lots of water, plant them there. Bananas like lots of nitrogen and potassium, so throw them lots of compost and banana peels. Wood ashes are good as well. Chicken manure is great for them, and they like chemical fertilizer too. You just have to give them water and fertility and they'll produce lots of bananas.

Each "tree" is just one shoot from an underground bulb. When a tree fruits, it's done and will not fruit again, so cut it down after harvest. Bananas and plantains are ready to harvest when the fruit fill out completely and start to turn dull green. If you notice a couple of fruit turning yellow in a bunch, the entire bunch can be harvested to ripen indoors.

Propagate bananas and plantains by dividing off little pups from existing clumps of trees.

### Cassava

Cassava is the best in-ground calorie bank you can grow in South Florida. Also known as yuca, the cassava plant is a staple root crop in most of the world's tropical nations. The plant is reproduced by planting woody stem cuttings from existing plants. Cassava isn't too picky about soil but produces faster if fed and watered. Poorly tended plants may make slightly bitter roots. Speaking of bitter, cassava roots must be cooked to be eaten. Boil them until soft and they are safe. If you don't cook them, they will poison you. This is no big deal, though, it's just a matter of proper preparation. Cook them and they're great. I've eaten lots and lots of cassava and I'm still here. Give cassava plants some space. 6' between plants is about right. About a year after planting you can do some exploratory digging around the base of a cassava plant to see if the roots have filled out. If you find a nice, fat root, go ahead and chop the top of the plant off and dig the entire thing up. Be careful not to destroy the woody stem portions, as you'll want those to plant and to give away to other gardeners for planting. There's no real harvest season for cassava. I've kept them in the ground for a couple years without harvesting, then dug them up and enjoyed the roots. I wouldn't go much longer than that because the roots can get woody, but you don't have to be super vigilant about harvest times. That's one reason why it's such a great survival crop—you can dig roots when you're hungry! Some varieties will produce harvestable roots at only six months or so but most of the cassava I've grown need at least nine months for a decent harvest.

Finally, the younger leaves can be boiled for twenty minutes and eaten as a green vegetable, much like cassava's cousin chaya.

Plant cassava canes in loose soil whenever you have them. I plant 12" cuttings half-way in the ground with the growth buds up. Don't point them upside down. You can also bury cassava canes on their sides about 2" beneath the soil surface. Shoots will emerge in a couple of weeks.

## Chaya

Chaya, also known as Mexican tree spinach, is a productive leaf vegetable with a good flavor. The variety with deeply lobed, sharp, jaggedy leaves isn't as good to eat as the type with more of a maple leaf shape. Chaya is a perennial green that starts readily from cuttings stuck in the ground. It makes a good hedge and has almost zero pest issues. Leaves should be boiled for twenty minutes to eliminate the natural cyanide content, then chaya is a hearty and healthy green. Young leaves are much better to eat than older ones. If the plants are well-watered and fed, the tops of new growth can be harvested and cooked as well. In the Caribbean I planted a partial hedge of chaya between my house and my neighbor's and we harvested the greens regularly during the rainy season. Cutting back chaya makes it branch and produce more leaves.

## Coconuts

Entire economies have been based around coconut palms. This valuable tree will keep you well-fed once you know how to utilize it. Younger nuts can be cut from the tree and chopped open for the sweet and refreshing coconut water inside. Then the pudding-like flesh can be scooped out and devoured. It is delicious! Older nuts can be harvested for both water and their nutmeats. The oil-rich meat of the coconut is a hearty food, filling and nutritious. Our Indian neighbors in South Florida would chop open mature nuts and let the meat inside dry out in the sun and peel away from the wood of the kernel, then they would use it in their cooking. If you have a lot of coconuts, you can extract the oil for cooking by shredding the meat and boiling it in water, then skimming off the resultant oil.

Coconut milk is not the water inside, though we used to call that "coconut milk" when we were kids. That's coconut water.

To make actual coconut milk, remove the husk from the coconut, and then crack open the hard nut inside. Rachel then uses a butterknife to pry the meat away from the hard wooden shell. Put the pieces of nutmeat into a blender and cover with water. Blend it thoroughly. Rachel then puts a clean, fine cloth over a colander in a bowl and pours the coconut slurry in the blender into the cloth, which she then picks up with the slurry inside and slowly squeezes, releasing fresh coconut milk into the bowl below. This milk is rich and excellent in curries.

Coconut husks are excellent additions to container garden soil mixes. Soak them in water for a week, then bury them in the bottom of pots, garden beds and container gardens for an in-ground water reservoir. Palm fronds are used for thatching and homemade hats as well as basket making. I also use palm fronds as a weed block, putting lots of them down on the ground and then stacking mulch on top. We also use them to shade garden transplants, gently laying the fronds over a newly planted bed. The heart of palm from a felled coconut palm can be harvested and eaten as a delicious vegetable.

On a hot day there is nothing like coconut water to keep you going—I used to skip a "proper" lunch on my tropical farm and drink coconut water instead. It quenches thirst and relieves hunger. A pole saw works very well for harvesting nuts from the trees.

Coconut trees benefit from having some seawater dumped at their bases. You can also just throw a few cups of sea salt around them occasionally.

Fully ripe brown coconuts sprout easily when buried half-way in the ground in a somewhat moist location. They take a few months to sprout and transplant readily.

## Dasheen/Taro/Eddoes/Malanga/Tannia

This is a complicated mess of species, so let's just call them "edible elephant ears." There are *Xanthosoma* called "dasheen" or "tannia" and *Colocasia* called "malanga" or "taro" or "dasheen." And sometimes you'll get something called "taro" that is obviously different from another elephant ear look-ing-thing called "taro." It's a mess, but just know if you find roots with these names for sale in a grocery or market, they are edible and easy to grow, whatever the Latin name may be. They grow well in Florida with zero pest problems and the edi-ble roots are good when well-cooked. There is a lot of oxalic acid in the roots so it is vital to cook them until soft or they'll burn your mouth and can close your throat up. I know some people who planted their entire Florida yard with a bunch of edible plants including peaches, loquat, yams, grapes, and dasheen. When the house was later purchased by non-garden-ers, they decided to dig up some of the edible roots the former homeowners had told them about. Unfortunately, the wife barely cooked the tubers and her poor husband taste-tested them. In moments his mouth, lips and tongue were burning and his throat was starting to close up. That was the last time they tried eating anything new from their yard, instead deciding to rip out all the "weird" stuff and just leave it grass. It's unfortunate because these edible elephant ears are a great staple when prepared properly, but knowledge of preparation is important!

Roots can be obtained from Publix and other supermarkets and planted in the late winter and early spring for a fall/winter harvest that same year. Divide clumps or roots and plant out to increase your stock. Edible elephant ears like lots of water and moderate fertility. As a bonus, they look lovely in the landscaping.

## Everglades Tomatoes

The Everglades tomato is Florida's easiest-to-grow tomato. It's the only one I recommend to normal gardeners, as growing tomatoes in Florida is usually an utter pain in the neck. Everglades tomatoes are tiny but sweet and abundant. They also self-seed and like to move into your landscaping. I would plant them in fall and late winter in South Florida. If you're lucky, they may live and produce through the entire year. The flavor is exceptional in sauces and is explosive when dehydrated. They are one of our favorite snacks in the garden.

Everglades tomatoes sometimes are harder to germinate than other types and may take a month or more to come up. One gardener reported planting them one year and not having them emerge until the next. As a wilder variety, this sort of behavior happens sometimes.

Everglades tomatoes like compost and full sun.

## Ginger and Turmeric

Both ginger and turmeric grow very easily in South Florida. They have a natural dormancy cycle, even in the tropics, so don't be alarmed when they appear to die late in the year. It's normal for the tops to die to the ground through the winter dry season. New shoots will emerge in spring. Ginger and turmeric like half-sun and moderately fertile soil. The plants are perennials and will make big clumps of plants if left unharvested. The leaves and roots are both good for a medicinal tea for upset stomach, queasiness, or morning sickness.

Plant the roots in winter and early spring for harvest the following winter when the leaves die back. Roots obtained from ethnic markets often grow just fine unless the buds have been removed.

## Green Onions

Green onions grow easily in Florida. Just plant a green onion from the store and it will happily multiply into a clump of green onions. You can cut the tops or dig entire plants for the roots. Green onions like fertile soil and some water in the dry season. They work well planted in landscape beds and under young fruit trees.

## Hot Peppers

Though bell peppers take some work in South Florida, most hot peppers are very easy to grow. If you're only grown hot peppers up north, you've never seen their full potential! Hot pepper plants can easily grow into 4' tall perennial shrubs in South Florida, often producing peppers for multiple years. Thai peppers and cayennes are both vigorous, with habaneros and jalapenos taking more work and usually living for a shorter period of time. Plant peppers with some compost around them and feed occasionally for better yields. Don't overwater them and make sure they get plenty of sun. We usually get way more hot peppers than we can eat or turn into hot sauce.

## Ivy Gourd

Ivy Gourd *(Coccinea grandis)* is an easy-to-grow Indian perennial cucumber. It's so easy, in fact, that it's been listed as an invasive species. I first met this vegetable thanks to our Indian neighbors who grew it on their chain-link fence. Thirty years later and the family is gone—but the cucumber vines are still there. Can you imagine any other cucumber surviving for that long? This cuke is a small-fruited species with somewhat tart fruits with thicker skins. Ripe fruits turn bright red and are inedible. Older green fruits may start to turn red inside and will not be good to eat. Pick the fruits when they are young and small. The vines make thick roots beneath the ground that allow the plant to regenerate if cut. Yes, it's definitely invasive, but it's a productive and tasty invasive that can produce fruit year-round. We turn the fruits into pickles by brining them in saltwater and letting them ferment on the counter for a few months. Start new plants from cuttings.

## Jicama

Jicama is a multi-purpose perennial. Every part of the plant is poisonous except for the root, which is a pleasant-flavored tuber. It is a nitrogen-fixing legume and the vines make a good green chop-and-drop plant, though you'll lose your root harvest that way. The seeds contain rotenone, which is a natural pesticide, though I have not tried to extract and use it.

The first time I grew jicama it was by accident. A good friend gave me seedlings which were allegedly from some seeds I had previously given him. I thought they were yard-long beans until they grew a little and I realized the leaves weren't right. The vines grew like crazy, covering the trellises I made and running off everywhere. Then they bloomed, and when I saw the flowers, I started to worry about them possibly being kudzu, which looked quite similar. Eventually I identified them as jicama and to this day I have no idea how we ended up mixing up jicama and yard-long beans. The seeds do not look similar, though the plants are both members of the *Fabaceae* family.

If you can't find jicama seeds, you can buy a root from the store and plant the whole thing in the spring (or any time, if you live in South Florida). It will grow vines which will then bloom and set pods, giving you plenty of seeds to plant the next year.

From seed, jicama roots take quite a while to develop. Mine made roots in the fall after being planted in the spring. Though the above-ground vines may die in one of our rare frosts, plants regrow readily from the roots. Older roots can get huge and become inedible. Harvest roots for the table when they are less than one year old.

The roots are edible raw or cooked. Plant jicama in a sunny part of your food forest or near something it can climb to reach the light.

## Kale

Kale grows easily from late fall through early spring. All types grow readily, including Lacinato and Siberian. A little hardier relative is called Ethiopian kale. It has a bit longer season. My friends Rick and Mart both hooked me up with seeds one year and we were quite impressed with how well they grew in the garden. It's a great green and grows to a solid 4" tall. Raw, Ethiopian kale has a mustardy kick to it that's a little zingy for my taste but still fits well into a salad. I really like it sauteed in home-rendered lard with scrambled eggs. Commenter "Organic Gardner" recently shared on my TheSurvivalGardner.com blog post about Ethiopian kale:

*I did an experiment and let several plants go to seed. I hung them upside down on my fence to dry and let the birds peck at the pods and spread the seeds. Now our entire fence is lined with the plant. I think technically it's a mustard green vs. a kale. Anyway it tastes great and keeps coming back every year and makes great edible and decorative landscape. The fence keeps it partially shaded/cooler so it doesn't bolt as quickly.*

I like plants that take care of themselves!

All kales are easy to grow and provide great health benefits. If you are allowed to keep chickens in your area, throw them a bundle of kale now and again and they'll scarf it down, then lay eggs with richer, darker yolks. We direct-seed kale in the ground and give it compost. It also benefits from foliar feeding. In hot weather, kale may bolt and turn bitter.

## Luffa and Angle Gourds

We first grew angle gourds many years ago, thinking it was going to make nice sponges as its cousin luffa does. We were wrong. Though it's a tasty vegetable, angle gourd has ribs and skin that stick tightly to the sponge within, even when well-dried. Luffa, however, is a dual-purpose crop, yielding useful sponges as well as being a very good vegetable. I grow both luffa and angle gourd, but if you were to pick one—luffa would be it. Angle gourds look really cool but are less useful.

The fruits of both species should be picked, peeled, and eaten when young. Older fruits become fibrous and/or bitter. If you harvest luffa or angle gourds and they are bitter, throw them in the compost. Do not eat them! My wife and I ate some bitter luffas once and it was like having the worst stomach flu, thanks to a toxin called cucurbitacin. This toxin isn't unique to luffa. It also occurs occasionally in cucumbers and summer squash, which are related species. It's not a big risk once you know it happens. Just pick your gourds when they are young and tender and if they're bitter, don't eat them.

Luffa and angle gourds love lots of space to climb. Give them wide spacing—at least 8' between vines—and they'll do better and be more productive than if you place them closely. Chain-link fences and small trees make good supports, as do cattle panel arches.

To harvest luffas for sponges, let the fruits dry on the vines and then bring them inside when they are brown so they can really get crunchy. Crush the fruits a bit by squeezing them in your hands and the outer skin will tear and fall off easily. We saw them into rounds and shake out the seeds from the inside. A single luffa or angle gourd contains enough seeds to cover your entire yard in vines.

Plant luffa any time during the year. They like the heat and the rain, however, so they'll do best when it's warmer out.

## Longevity Spinach

Longevity spinach (*Gynura procumbens*) was made for

Florida. It's so easy to grow and healthy that you'll wonder how you went without it. The leaves have a fresh, earthy green flavor that I immediately liked. It's a good salad stuffer or a cooked green.

Longevity spinach is a half-vining herbaceous perennial that tolerates poor soil, drought and heat but benefits from water and compost. Propagation is by stem cuttings. Just break off a piece of stem, stick it in the ground, water it and it will root. Longevity spinach takes sun and shade well. Its pretty purple-and-green cousin Okinawa spinach has a milder flavor but was much less hardy in my North Florida garden, wilting in the heat and dying in the winters. It should do better in South Florida, though I bet it prefers some shade.

Longevity spinach does not like frost and often dies if it freezes to the ground. If a frost is coming, take a few cuttings from your plants in fall and put them in a pot to root, then plant them out again in spring.

Finding longevity spinach may not be easy. Ask around and make friends. People are often trading plants in the comments of my YouTube videos, so be friendly and you'll soon find yourself with some cuttings—and once you have them, you're off to the longevity spinach races.

## Mango

Mangoes may be the best fruit in the world. Almost everyone who has tried fresh mangoes wants to grow a tree.

Mangoes grow excellently in South Florida. The tree

growing in the Great South Florida Food Forest has been bearing heavily for years, and we've cut it way back multiple times. It just keeps making fruit!

If you let mango trees grow to their full height, they can be monstrous. I've seen them as tall as forest oaks in the Ft. Lauderdale area, at least 60 feet above the ground. This makes harvesting difficult but they are really striking trees. Fortunately, mangoes can be pruned a lot smaller for little spaces, so long as you keep cutting them back. Mango wood is very beautiful and has been made into furniture and musical instruments. Malaysian luthier Jeffrey Yong is famous for his beautiful mango wood guitars. Tired of importing wood from Europe and South America, he decided to experiment with local tropical woods and discovered that mango is not only lovely but has a remarkable tone. If you ever see a mango felled by a hurricane, see if you can save some of the wood.

Fresh mango seeds are easy to germinate if the husk is cut away and the embryo inside planted. Seeds with more than one embryo produce true to type, but mono-embryonic seeds are wildcards and may produce fruit that is better or worse than the parent tree, depending on the genetic lottery. If you want an exact cultivar, graft scion wood from a known type onto a seedling tree. You can also air-layer mangoes, though it doesn't always work. I attempted to air-layer my grandpa's excellent mango tree years ago without success. Last year I tried again with another tree and it worked just fine—and bloomed the first year. Try, try again.

## Moringa

Moringa, "the miracle tree," is touted as a cure-all and has now been planted in yards all over Florida.

Moringa is a fast-growing tropical tree species with multiple uses. In its native range, the tree's large pods are picked young and served as a delicious vegetable comparable to asparagus. Here the moringa is often grown for its tiny edible and medicinal leaves. Because of the tree's remarkable ability to mine the ground for nutrients, the leaves are loaded with nutrition – and even contain complete protein, a relative rarity in the Vegetable Kingdom. There are claims that the tree also kills fungal infections, fights cancer, gives you the ability to fly, etc. I'm not sure about all those bits and pieces, but its nutrition has been proven in the lab and on the ground in Africa, where dried leaves are used as a powerful antidote to malnutrition in infants and nursing mothers.

Beyond those benefits, moringa also grows at a ridiculous rate. The first time I planted seeds, the trees shot towards the sky at an astounding speed, reaching 20' before winter frosts knocked them back to the ground. This rate of growth means you'll have plenty of leaves to harvest. Bonus: moringa leaves are excellent livestock feed. The tree is also good for chopping up and adding to compost piles, since its soft wood deteriorates rapidly. It's a staple chop-and-drop tree in my tropical food forest projects. I've also dried the thin leaves in my greenhouse and then crushed them into powder. I then sprinkle that dust over newly prepared garden beds for a little extra dose of fertilization. Moringa seems to give young plants a kick.

Moringa is propagated from seeds and cuttings, though the first method gives you much stronger plants. Cuttings sometimes take and sometimes don't. If you want to give cuttings a try, lop off a branch ranging in diameter from 1-2" and at least 2' long and bury the bottom third in the ground. It usually starts sprouting new growth in a few weeks—or it decides to rot. I might use cuttings for an instant barrier fence, but seedling moringa trees can reach 20' tall in their first year so I usually just grow those. I don't like the weak root system on cuttings.

Beyond being good fertilizer and livestock fodder, moringa leaves are nice added to soups, salads, stir-fries, and eggs. Snapping the large, compound leaves off the tree is easy, and once you do that, you can strip the little leaflets off into whatever you're cooking. This plant is a nutritional powerhouse. A little moringa each day keeps the doctor away.

Plant moringa in full sun.

## Pineapples

Pineapples grow easily in South Florida. Pineapples have shallow root systems and can take a couple years to produce from a top or a slip, yet you can start them any time you want, put them aside, start more, put them aside, and eventually, you'll have tons of pineapples. Just do it in between taking care of your faster-producing plants and you'll get there. My grandpa planted dozens of pineapple tops in his Ft. Lauder-dale landscaping and they bore delicious fruit every year with almost no care.

In South Florida, pineapples can be tucked into any empty space as you have planting material. The thorny types are particularly good for boundary lines and places where you don't want traffic.

Pineapples like mulch and compost. If the leaves start to get very yellow or reddish, it often means they're starving. A light fertilizer solution or some compost will make them happy again.

As your plants grow, they often make side shoots which you can separate and plant. Some pineapple varieties make a bunch of small shoots on the base of the fruits which work as planting material.

When planting pineapple tops, peel off the remaining fruit and the bottom few leaves of the top until you see some little root nubs beneath, then plant the top in the soil. This helps the plants get established faster. I do not recommend rooting tops in water as they'll often rot. Plus, it's easy to just stick them in the ground to root.

## Seminole Pumpkins

In Florida, the winning pumpkin/winter squash is the native Seminole Pumpkin. It is delicious when roasted with butter and salt. It also makes excellent pies. Though its origins are murky, it was reported as a native crop by the Spanish back in the 1500s, so if any of us can be called natives, the Seminole pumpkin surely can. The Seminole pumpkin comes in a wide range of varieties, likely due to crosses. Some years ago, I set up an online gallery of Seminole pumpkins to catalog strains and growing locations sent to me by readers. There are some that ripen green, though most ripen tan. There are varieties clocking in at 12lbs or more and others that are only a couple of pounds. What connects them is their rampant growth and reliability, their resistance to diseases and vine borers and their rich orange sweet-flavored flesh and long storage time. Chances are the variety has drifted and crossed with other *C. moschata* species, as it's more of a land race than a true variety, but boy oh boy, it's a good grower and a top-notch survival crop. It isn't the only pumpkin/winter squash that thrives in Florida, however. There are many varieties of "calabaza" from Central America and Mexico that will take the heat and produce well. There are also varieties from farther north, such as the "Tan Cheese" pumpkin which do well in Florida. Most northern varieties of pumpkin suffer in Florida due to powdery mildew, vine borers and the heat, so don't get carried away looking at the gorgeous varieties from cooler climes. Hubbards failed for me in Florida, even though they loved my garden in Tennessee. Boston Marrow is probably a no-go, as are most of the *C. pepo* and *C. maxima* varieties I tried. Butternut, being a *C. moschata*, can do well in Florida but is regularly eclipsed in production and disease resistance by Seminole Pumpkins. Seminoles just don't quit. If you plant one pumpkin/winter squash, make it Seminole. Many gardeners in the state save and share seeds. It's good to have friends. If you don't, you might have luck finding Seminole pumpkins in some seed catalogs. I usually avoid ebay as a seed source as there are too many scammers on there, but there are some newer seed shops on Etsy that carry good varieties, including my daughter's "Good Gardens" store at www.etsy.com/shop/GoodGardens.

In South Florida, you can plant pumpkins and winter squash year-round. Direct seeding is better than transplanting for vigorous plants.

Pumpkins and winter squash need lots of room to roam. They can be trellised but that is a risky business as they like to root at the nodes of the vines and gain more strength that way. If they climb, they cannot root all over the place, increasing their chances of being murdered by Pumpkin Public Enemy #1, the vine borer. A vine that gets drilled by vine borers but has a lot of rooted sections will shrug off the damage and keep going. Sometimes the damaged vines will even end up divided into two separate plants that will both go on to produce pumpkins. The modular nature of *C. moschata* pumpkins really helps them cope with the stress of Florida growing.

Plant pumpkins and winter squash on mounds that are 8' apart in all directions. You can probably push them as close as 5', but don't go any closer than that as crowding reduces yields. Put them in full sun. They'll take half, but tend to run for the light if they get the chance and will be weak in shade. The vines can grow a surprisingly long distance and will fight for space and reduce yields if you crowd them. I grew pumpkins along the edge of my tropical garden where they could ramble down into the drainage ditch and around the back of my compost pile and under my starfruit tree, which kept them from running wild all over my tomatoes and pak choi. At my current location I grow them in their own patch of dirt well away from the main gardens.

I grow pumpkins on mounds, like most everyone, because that's just the way everybody does it. Where I change the game is by burying a bunch of nitrogenous material under the mound. I call it the "Melon Pit" method, because it's also great for growing melons.

To make melon pits, dig a hole that's at least 2' deep. 3' is better. This is easy in sand but hard in clay or on lime rock soils. Do your best. Now throw in some kitchen scraps and other horrid stinky things. Meat or fish scraps are great. Dog droppings are fine. Humanure, goat organs, an old lasagna, raw chicken manure, hair, a shovel of compost, a dead pet, your enemies, whatever. We grew a great Seminole pumpkin over a dead rat once. The idea is to put some super rich food underneath the vines that the roots will find and kick-start the plant's growth. This will not hurt you—no one will know what you did and it won't make you sick. Once you've thrown a horrifying thing or three into your melon pit, cover it over and make a mound on top. Make sure your loathsome, wretched, vomit-inducing, nasty, horrible material is a solid 2-3' down if you can, as that usually keeps roving critters from digging it up again. Now plant 3-5 pumpkin seeds in the mound around an inch deep and water deeply. In 4-8 days, they should pop up. Within a week or two, their roots will go deep enough to hit the scary stuff and they'll turn deep green. In a couple more weeks they start running and you are off to the races.

Keep the weeds and grass down at first so your pumpkins can conquer the space without competition. Once they really get rolling, you won't have to worry about weeding as the vines will cover everything. Water as needed. It's normal for pumpkins to wilt in the midday heat, so don't worry too much about that, but if you get up early in the morning and your pumpkins are wilting, you either need to water or look for vine borer damage. Vine borers are nasty insects that lay their eggs in the stem of pumpkin vines. The eggs hatch and turn into worms that chew their way along inside the stems, causing your plants to lose their connection to water. As I mentioned above, Seminole pumpkins can often shrug off the damage due to their ability to root at the nodes, but I've still lost them on occasion. I skip the evil pesticides and just let them root everywhere. If you move the end of pumpkin vines that are more than a few days old, you'll see that the nodes on the new growth are already starting to put out roots. I encourage this by throwing handfuls of sand and mulch over nodes in the vines, figuring that more roots equals a much higher chance for the plant to survive the inevitable borer attacks.

If white mildew shows up on your pumpkin leaves, try treating it with a few tablespoons of plain yogurt shaken up into a spray bottle of rain or well water (chlorinated water kills the bacteria in yogurt!) and sprayed on the leaves in the evening. The beneficial bacteria in the yogurt seem to beat the infection down. Generally, I only have mildew appearing on my vines late in the season when they're about to die anyhow so I don't worry about it.

Once you have fertilized flowers, female pumpkin blooms drop their yellow flower and begin to swell. They start out a pale green color and mature to their final colors over a month or so. A pumpkin or winter squash is ready to harvest when the stem yellows or browns or the main vine dies. You'll be tempted to cook one right away, but the flavor of a newly harvested pumpkin pales in comparison to one that has been allowed to sit on the shelf and cure for a month or more. That's when the sweetness and flavor really develops inside. It's wonderful to see a row of pumpkins sitting on a shelf as well. Seminoles can keep for quite a long time. The smaller ones with the darker tan skin seem to keep longer than the larger ones with paler tan skin, but I have had both sitting at room temperature inside for over a year without spoiling. One small one I was given by Jacksonville Permaculture Guru Alex Ojeda kept for two years before we opened it up and ate it. It was rather dry inside but still sweet and edible. What a great survival crop! My guess is that the Indians inadvertently bred this variety for storage by eating their pumpkins through the year and saving the last ones for seed, though that's just conjecture on my part.

Pumpkins can be added to soups, cut in half with the seed mass scooped out and roasted in the oven or baked into delicious pies. The taste of a Seminole pumpkin is much like that of a good orange sweet potato, rich and buttery. The seeds can also be roasted or fried in oil for a delicious snack. Just be sure to save some for next year. *C. moschata* pumpkin types regularly cross with others inside the species, which means if you grow a long-necked Central American pumpkin next to your Seminoles, then save seed, you may end up with weird, long-necked Seminole types the next year.

As a final note on pumpkins and squash—they do not like the high heat and bugs of midsummer in Florida. Get them in early. One year I had them pop up in my compost heap and run rampant, then die back in July and August, then somewhat recover in the fall and bear me a second crop. This is not common, however, so plant accordingly.

In a food forest, I find pumpkins and winter squash to be best utilized in the early phases when you have lots of open space. They make a great ground cover in sunny spots but do not like much shade. If you just cleared ground and started a food forest, consider putting in some pumpkin hills in between your newly planted trees so you can get yields while you're waiting. As the food forest matures, pumpkins must be relegated to the edges where they can get the sun they need to thrive.

Save seeds from fully ripe pumpkins by washing them and drying them out well, then storing in the fridge in a sealed bag or jar until needed.

## Starfruit

Starfruit is one of my favorite fruits.

There is a beautiful tree growing in The Great South Florida Food Forest Project which has provided baskets and baskets of fruit over the years.

Starfruit, also known as carambola, usually bears two crops a year in South Florida.

If you've only had starfruit from the grocery store, you haven't tasted starfruit. Those bland and watery things are terrible. Fresh starfruit has a juicy sweet-tart tropical goodness that's very refreshing. Don't eat too many if you are subject to kidney stones, as they are high in oxalic acid.

The trees do not get very big. They spread sideways and have attractive bark and feathery leaves. Some years ago I stood transfixed beneath a 15' tree in a friend's yard, looking up towards the sky. The sight of the semi-translucent fruit hanging like Chinese lanterns in the tropical sun was transcendently beautiful. I could have just sat down and stayed there for hours, staring up through the branches.

You can propagate starfruit via seeds or grafting. Some

seedling types taste sour or have bitter skin, though some are very good.

## Sweet Potatoes

Sweet potatoes are a top-notch staple calorie crop in the Sunshine State. Sweet potatoes are planted by cutting vines from existing sweet potatoes and planting them out. These are called "slips." In my garden I plant sweet potato slips about 12-16" apart in rows 3' apart. Loosen the ground before you plant them to ensure they can make decent roots. In the food forest, I tuck them here and there around trees and perennials in places where I have at least half sun. Slips do not need to be rooted before you plant them. Just cut pieces of sweet potato stem, remove all the leaves except for one or just the little ones, then plant them on their sides a couple inches deep with one end of the slip sticking out of the ground, then water them. For a few days, they'll wilt and look awful, but they root readily and start running quickly. At the base of each planted slip, potatoes will form in a few months.

If you do not have slips to plant, it is easy to start your own from sweet potato roots. I once bought an assortment of different sweet potatoes in varying colors from an organic market, started slips, then planted them willy-nilly in my garden. That year we had a lot of fun digging potatoes that ranged from white to yellow to orange and deep purple. To start slips from sweet potato roots, lay them on their sides in soil just beneath the surface. Water regularly but don't make them sopping wet. Keep this flat in half-sun. In a few weeks, shoots will start emerging from the roots. As they grow to 12" long, you can carefully trim or break them off to plant out. Planting season for sweet potatoes is year-round in South Florida. Sweet potatoes will grow in mulch or bare ground but they do not like full shade. Dappled shade will still get you some potatoes and I had good luck planting them in mulched areas of my food forest in North Florida, where they bore quite well. It's best to plant in different areas from year to year, but you might get away with growing them a couple of years in the same spot. My potato yields decreased greatly when I just let the vines keep running and running for a couple of years.

Sweet potatoes like organic matter in the ground and really appreciate potassium (wood ashes are a good source) but do not feed them much—if any—nitrogen or they will produce abundant vines and few roots. I have made this mistake before. The plants looked amazing but made almost no roots whatsoever.

Sweet potato leaves are edible raw or cooked but are reportedly much better for you cooked as they contain some anti-nutrients when raw.

After a few months in the ground, you can do some digging to see if your plants are making roots. The boniato types I have grown—that is, mostly red types with white flesh from Latin America and the Caribbean—took much longer to produce than the traditional American orange types. White varieties are starchier and drier and not as sweet but make a nice change from the very sweet orange types.

Sweet potato roots can be boiled or fried and, like the leaves, are not good to eat raw.

When you harvest your sweet potatoes, give them a week or two to lay out and cure in a shady, dry location or they will not taste as nice and sweet as the ones you buy from the store. Do not wash them before curing and be careful not to damage roots when you dig them or they won't keep as long. Once they are mature in the ground, you can dig them as you like over the course of some months or just harvest them all, cure them, and let them sit in a cool, dry location in cardboard boxes or baskets. They keep in my air-conditioned pantry for months. Remove any potatoes that go bad and eat damaged ones first.

Sweet potatoes make a very good ground cover for new food forest projects as they'll run rampant and suppress weeds while still getting you a harvest.

## Yams

As everyone knows by this point, I am a huge fan of true yams, also known as "name" yams (said "nay-may"). They are not related even distantly to sweet potatoes despite sharing the name "yam." Yams are a climbing vine that can make huge roots beneath the ground. So far, I have grown purple, white, and yellow greater yams (*Dioscorea alata*), potato yams (*Dioscorea escuelenta*), Chinese yams, also known as cinnamon vine (*Dioscorea polystachya*), the edible variety of *D. bulbifera*, which bears potato-sized roots on hanging vines and is quite rare, Lisbon yams (an improved *D. alata* cultivar) and a few other types I can't find the names for. My favorite producers are the white/yellow *D. alata* types, also known as the "greater yam" or the "winged yam." It's on the Florida invasive species list which tells you just how easy a crop it is to grow. In the Caribbean and tropical portions of Asia, Central America and Africa, yams are staple crops that feed everyone through the fall and winter months and on into spring. On some islands, they grow rampantly in the wild and are dug with machetes by the poor during the winter dry season. In Florida, finding propagative material can be a wild goose chase but it is by no means impossible. I regularly see "name" yams for sale at Publix and if you buy one of them, you can divide it and plant multiple hills of yams. There are also quite a few edible yams growing wild in the state; however, make sure you do not mistake the common air potato (*Dioscorea bulbifera*) for the edible *Dioscorea alata*. The wild specimens of *D. bulbifera* in Florida are not edible and are poisonous. *D. alata*, however, can be found in the wild and propagated provided you can tell it apart from its poisonous cousin. You can also look up my yam/air potato identification video on YouTube to see the difference in growth and leaves and you'll be an expert in no time.

Yams can be propagated by a variety of means. You can take existing roots, like the one you found at a produce market, and cut it into pieces about half the size of your fist, making sure you keep as much of the skin intact on each piece as possible, then dust the cut portions of the individual pieces with ashes to prevent rot and fungi, then plant them in the ground or in a pot to sprout, then transplant those sprouted pieces into the ground later.

I have also started *D. alata* from cuttings, but did not get to compare the yields to tuber-grown and I am sure they will not produce as large a root in the first year. One of the best ways to grow most yams is from the small, bulbous, aerial roots that grow on the vines in fall. These are called bulbils. As the yam vines die back in fall and winter, they fall to the ground and sprout in the spring.

Yams can be planted from fall through spring. They have a pronounced dormancy period that lasts from November until some time in the spring, earlier or later depending on the species and rainfall. Yams like compost and mulch at planting but don't seem to need much else. And they don't even need that. Plant your yams about 2' apart in rows 4' apart, or make individual hills a few feet apart. Plant cut pieces, bulbils or sprouted pieces 2-4" inches deep so they don't dry out in the ground. It's important to make sure you plant in a spot with loose soil so the roots can expand down. Hills aren't a bad idea because they're easier to dig later, though they dry out faster than flat-planting. Make sure you have a trellis or a tree in mind before the vines start growing out of control. When they come up, they shoot up very fast, probably growing a foot a day or more, and will turn into a tangled mess if you aren't ready for them. I have had good luck letting them climb along fences and up into trees. It may be shady under a tree but they'll climb up into the canopy quickly and unfurl an abundance of leaves at the top, capturing the sunlight they need to grow their tubers. If you have a wide-spreading tree, you can plant yams under it and run strings down from the branches and let them climb up into it. I do not recommend covering all your favorite fruit trees with yams but I view all poincianas, oaks, and other non-food trees to be fair game as living yam trellises.

When the vines die back in fall and winter, you can dig yams or leave them in the ground until you want them. Roots usually run from 1-6lbs in size the first year. Their size depends on how well they were treated, how much rain they got and how fertile the soil is. If you wait too long to dig yam roots they'll start sprouting again in the spring. You can dig them when the vines first appear but don't wait more than a couple weeks after vines emerge as the roots start deteriorating as they feed the growing vines. If you want to grow really big yams or don't need the food at the time of re-sprouting, just let the vines go. They'll suck the energy out of the yam in the ground and it will shrink and rot away as it feeds the new aboveground growth. The second-year growth on well-rooted yams is incredible. I've seen a fat vine shoot up eight feet from a large root without putting out a single leaf, reaching for the sky and sunlight as fast as it could grow. These vines will be bigger and make more leaves than they did in the first year. Over summer and fall they'll build a new and larger root beneath the ground. Second-year yam tubers can easily surpass 20lbs.

Once you have yams, it's easy to grow more. You'll get a few bulbils for planting from first-year yams and a bunch from second-year ones. So long as you don't dig them, they just keep growing every year, though parts of the roots will get gnarled and woody. One really cool thing about yams is that you can harvest them in fall, then cut off the top couple of inches of the root where the vine had been, then plant it back into the hole where you dug the yam. This top part is called the "head" and can be planted over and over again and it will grow new vines and a root beneath it. If you want to keep already dug yams over the winter, just put a bunch in a bucket somewhere and throw some leaves or sawdust or a bit of soil over them until you want to plant them again in spring.

Most yams contain oxalate acid crystals in the roots so they cannot be eaten raw, with the exception of Chinese yam. Some people's skin is irritated by peeling yams so wear gloves when you process them. Just cut off the top couple of woody inches—the head—and then peel the rest of the yam. It's slippery and slimy when raw so be careful not to slip and cut yourself as you peel. Then chop the peeled roots into pieces and use them as you would white potatoes, cooking until they are fork tender. This usually takes about twenty minutes of boiling. They are good in stews or boiled and mashed. Yams also cook quite well in a crockpot. Mashed yams with cheese, butter and salt are excellent. If you cut yams into smaller pieces and boil them until they are soft, you can then pour off the water and fry them into yam fries. The flavor of most yams is similar to a white potato, starchy and not sweet. If you make beef stew and substitute yams for potatoes your guests won't even know the difference. Though slimy when raw, they are not at all slimy when cooked, so fear not. This is just a great staple carbohydrate crop. I am grateful to my friend Craig Hepworth for sharing his love of yams with me years ago, when he told me how he thought it was the top survival staple crop to grow in Florida. After my first year of growing them, I found myself in complete agreement.

If I had to pick one staple survival crop for Florida, yams would be it. They are a must-add carbohydrate crop for any serious backyard garden, and once established will often keep planting themselves without any help from you.

## Yard-Long Beans

The best green bean I have grown in Florida is the "yardlong bean," also known as the "snake bean." They are a very productive and heat tolerant bean when picked green, though the dry beans are thin and not good for much other than planting. You can plant them at any time of the year. Grow them on a sturdy trellis and they'll produce continuously for a month or more. Let some pods dry on the vine and you'll have seed for the next round. This has been my main green bean for many years now and it doesn't let us down. Just watch out for ants, as the blooms seem to attract fire ants.

## Mulberries

This low-maintenance and easy-to-grow tree will produce more sweet fruit than you can handle. If you have a small yard, no worries—just prune them back after they fruit. You can cut the living daylights out of mulberries and they'll regrow. Mulberry leaves are a good animal forage and the branches can be used to make biochar. Try to avoid planting purple-fruited types near driveways or areas where you're likely to step on them and track purple juice into your house and onto the carpets. They drop a lot of fruit all at once. There are also white-fruited varieties which do not stain and can be planted wherever you like. White fruits taste like honey and are quite pleasant. The common Dwarf Everbearing type is a little slow to get going and does not make the best fruit, though it bears abundantly and is easy to control in a small space. We have one growing well in the Great South Florida Food Forest Project along with a 5th Street mulberry. Unfortunately, the latter does not fruit much at all in South Florida. The best variety we've encountered for the entirety of the state is one I named "Rachel Goodman," after my wife. This variety is propagated exclusively by Scrubland Farmz nursery north of Ocala. The flavor is excellent and the tree is attractive. The mother tree is from Ft. Lauderdale and grew on my wife's street when she was a little girl. Later it was destroyed in a hurricane but cuttings were saved by gardener Guy Seligman, who started multiple new trees in a lot across the street from the location of the original mulberry. When Rachel and I went looking for the tree we had visited as children, we saw it was gone—and then met Guy, who showed us the children of that original tree and shared cuttings. Great save! It's a nice tree, not too large, with an umbrella-like canopy.

Seedling mulberries can take 10 or more years to bloom and come in both male and female forms. Males only produce pollen; females only produce fruit. However, pollination is not required for fruiting—and without a male, you get seedless fruit—so there is no need to keep male mulberries around. Unless you're trying to breed a new variety, it makes much more sense to start mulberries from cuttings or grafting than from seed. If vegetatively propagated, mulberries will often bear in their first year after planting.

# CHAPTER 7

# South Florida Gardening Tools

I admit it: I like tools, especially hand tools and antiques.

But you don't really need much to work a backyard garden in South Florida.

You don't need many tools to take care of a garden, particularly with South Florida's easy-to-work sand.

There's no need for a tiller, as the sand is already loose and grass and weeds are easy to remove.

There are only a few tools I need to use regularly, and a few more I like. We'll run through them quickly and I'll share their uses and what you should have on hand for easy gardening.

## 1. A Machete

A machete can be used for chopping coconuts, taking limbs off trees, digging holes, clearing brush, harvesting bananas and, in a pinch, slicing mangoes. I don't even use a transplanting trowel anymore because a machete works great for popping in seedlings. When you want to chop up materials for the compost, just use your machete. I lay materials on the ground and chop through them right into the sand. Sharpen it with a good file every time you use it and the blade will work for you. Tramontina makes great machetes if you can find them. Martindale is another good choice, but is hard to find in the United States. Shorter machetes are easier to wield and work for most jobs. The standard Latin American shape is the best for most jobs, though cane machetes have a squared-off end with a hook on the back that is good for pulling down fruit and pulling in tall grass and sugarcane. Most hardware store machetes are fine, too, though I do not like the weak metal on the Gerber "Gator" machetes and avoid those after having one ding up on me during the first use. Be careful with machetes, though. I cut through the back of my hand with one once and required surgery to fix two tendons. If you're tired, stop. If you get distracted, you can hurt yourself fast.

## 2. A Spading Fork

Spading forks allow you to loosen the soil and get compost worked into the ground. We use one when making new beds and to loosen up grass and weeds for easier pulling. Cheap models have thin, flat tines that bend easily, so make sure you get one with some strength.

### 3. A Shovel

A long-handled combination shovel is good for digging holes, though South Florida sand is so easy to dig that I've also used a flat-ended shovel without issue. If you have roots in your soil, a sharp serrated type like the Root Assassin or the Root Slayer shovel is a nice addition to your digging arsenal. That allows you to slice right down through roots without having to dig around them and chop with a hatchet or a machete. They also work well for dividing banana pups off a clump of banana trees.

### 4. Loppers and Hand Shears

A pair of long-handled loppers is excellent for taking off tree branches. The ones without adjustable handles are easier to use and less likely to slip around during use. Small hand-pruners are good for tighter work and grafting and pruning shrubs and trees, as well as taking cuttings. Felco makes excellent ones, but the little cast-metal types in the hardware store are much cheaper and also work well.

### 5. Pruning Saws

A sharp hand-held pruning saw makes short work of smaller branches. Bow saws are better for larger limbs. A long pruning saw on a pole is good for tall trees and for harvesting high-up fruit and coconuts. We use ours regularly in my Mom's South Florida food forest for taking down coconuts and keeping the mango tree under control.

### 6. A Long-Handled Fruit Picker

If you have tall fruit trees, it makes sense to buy a long-handled fruit picker. These are basically a little basket on a pole with some hooked tines at one end that are great for picking ackee, mangoes, citrus, and other tree fruits.

### 7. Watering Cans and Pump Sprayers

We buy inexpensive 2-gallon plastic watering cans to water with compost tea and liquid fertilizer as well as for irrigating transplant flats and small garden spaces that aren't near a hose. A pump sprayer can also be used to foliar feed. Just make sure there isn't any debris in the container as that will quickly clog the tiny spray nozzle. I find the backpack types to be unwieldy and prefer the simple ones you carry by a handle in one hand while spraying with the other.

### 8. Garden Hoes

A standard garden hoe, when sharpened and adjusted to the right angle, will decapitate weeds in sand with ease. I sharpen ours with a file before weeding and it makes the job easier. That way you don't have to chop hard. You just slice through the weeds easily, cleaning up a space in no time. Another option is the scuffle hoe, which has an oscillating blade that cuts on the push and the pull. It's very easy to use in sand and is faster than a regular hoe in most conditions. A standard hoe is more versatile, however, as you can use it to drag soil around and to make furrows. We also often lay the entire hoe handle on freshly loosened soil and step on it to make seed-planting furrows. Another type of hoe, the "grub" hoe, has a broader and heavier head which can be used to

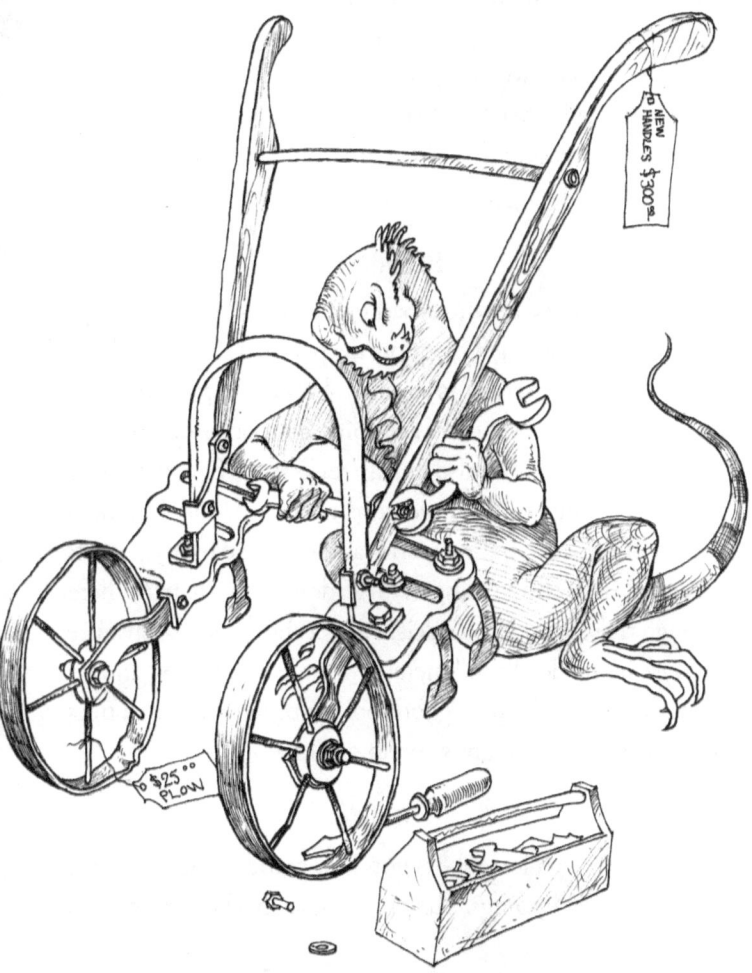

dig beds, remove weeds, and even make transplant holes. I like the various larger hoes sold by EasyDigging.com. If you have a larger backyard row garden, a wheel hoe is very good for cultivating, making furrows and weeding, depending on what attachments you have. They are much more expensive but speed up weeding in a big traditional garden. I have some vintage Planet Jr. wheel hoes I use regularly. You can also get very good wheel hoes from Hoss Tools and other suppliers.

## 9. Broadfork

The Meadow Creature broadfork is a tough tool that helps you loosen soil and get compost deeper into the ground. It's also a good replacement for a tiller, should you need that. It's not a must-have, but if you are doing a lot of gardening they are quite useful. They'll also pull up rocks if you have limestone boulders in your yard, or old construction debris. I found a bunch of buried stepping stones and concrete in my parents' backyard when I was using my Meadow Creature back there. Though there are other models, I prefer the Meadow Creature because it's made of solid metal, it's made in the U.S., and it's near indestructible. It's a good tool for pulling up small trees and shrubs, too, as it gives you lots of leverage.

## 10. Rakes

A metal-tined leaf rake is good for gathering leaves, prunings and grass clippings. The plastic ones are cheaper and also work. A hard-tined metal landscape rake is my favorite tool for making fine seed beds and shaping garden beds.

## 11. Power Tools

Bagging lawnmowers are useful for cutting grass and shredding leaves to add to the compost pile. A good string trimmer is also good for getting your yard in shape and cutting down tall grass to put in gardens. Some people prefer to use a scythe, but there is an art to using this ancient hand tool and you may or may not like one. Chainsaws can be useful—especially after hurricanes—but are touchy and usually can be replaced with a bow saw in smaller yards. A friend of mine loves his electric chainsaw but I have not used one. Instead, I have a smoke-belching Stihl that roars and tears through oak while making the neighbors nervous.

## 12. Pitchfork

Don't try to use a pitchfork for loosening the soil. A gardener recently sent me a picture of one he had broken while trying to make a garden bed. That's what a spading fork is for. A pitchfork is for throwing mulch around and turning your compost, should you choose to do so.

That's pretty much the set of tools I use in a South Florida Garden. You could probably just garden with a rake, a machete and a shovel, but all these tools are useful in certain applications. Buy decent tools and they'll last a long time. The sand is not hard on tools like rocks and clay can be.

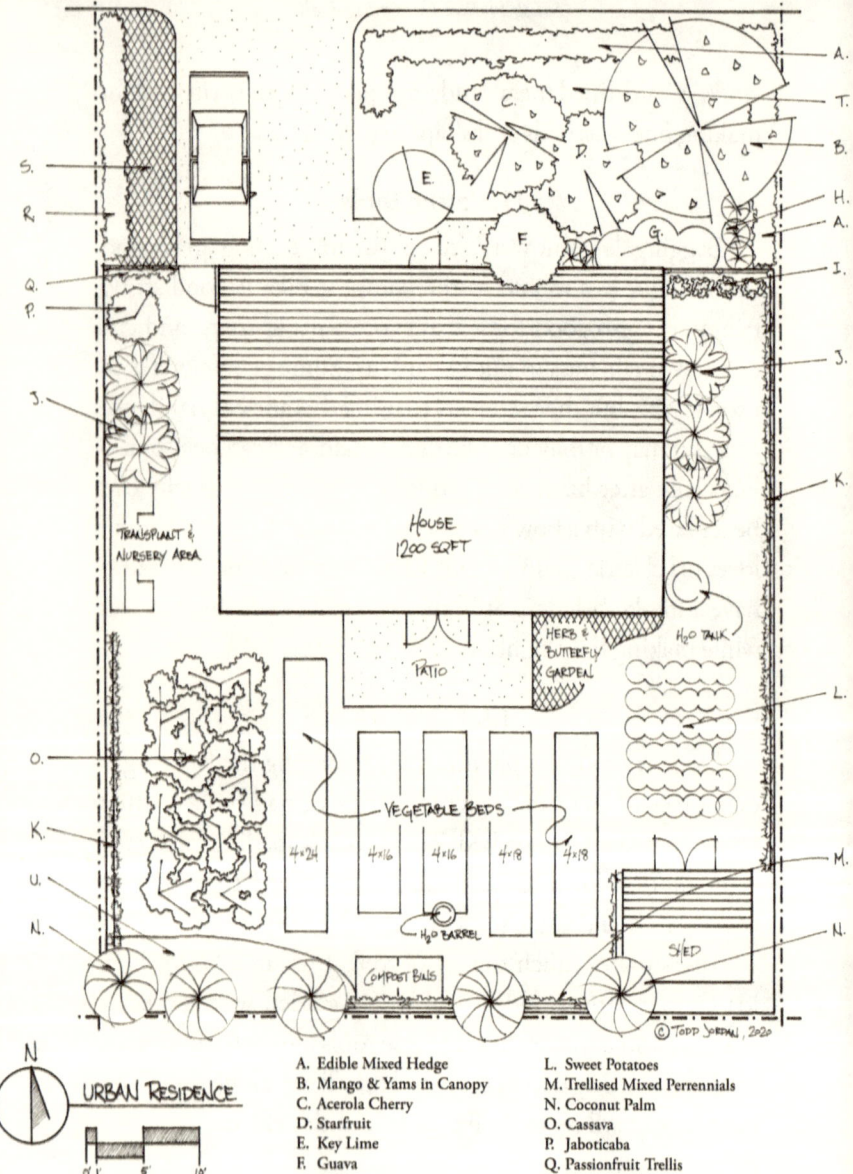

A. Edible Mixed Hedge
B. Mango & Yams in Canopy
C. Acerola Cherry
D. Starfruit
E. Key Lime
F. Guava
G. Surinam Cherry
H. Pineapple

L. Sweet Potatoes
M. Trellised Mixed Perrennials
N. Coconut Palm
O. Cassava
P. Jaboticaba
Q. Passionfruit Trellis
R. Edible Chaya Hedge
S. Edible Landscaping

URBAN RESIDENCE

0' 1'     5'     10'

© TODD JORDAN, 2020

# CHAPTER 8

# A South Florida Backyard Gardening Plan

B ack in 2020 I wrote a book titled *Florida Survival Gardening*, in which I covered how to grow food in Florida even in tough times. Unlike this little book, *Florida Survival Gardening* covers the entire state. In it, I included a plan for a South Florida survival garden. I am reprinting it here so you can get an idea of just what is possible, even in a small lot. Our tropical climate is an incredible gift.

## A Survival Garden Plan
## for a Small South Florida Yard

This garden plan is for a small suburban South Florida home on a 1/10 acre lot. Top is North. The lot is 70' x 90' with a 1200 square foot home in the middle.

If the front yard is kept well-maintained, you can generally get away with whatever you like in your backyard, provided you have a fence. I highly recommend a privacy fence around the backyard if you can afford it. A wooden fence should cost less than $3,000 and will spare you from prying eyes and calls to code enforcement. It also makes your garden feel more like a private oasis from the madness of the world.

The trees in the front yard can be pruned to keep them small, or you can plant less and let them run up to full size. I recommend topping mango trees when small and letting them make multiple side trunks. Otherwise they will get gigantic and drop fruit from a mile up. The edges of tree canopies can overlap a little but you shouldn't let large fruit trees completely overshadow smaller fruit trees or else the smaller tree is unlikely to bear much—if any—fruit. A chaya hedge is quite serviceable but if you'd like something prettier, cattley guavas, pineapple guava, cocoplum or Surinam cherries will fill in nicely and produce edible fruit. If you mulch the front yard and plant sweet potato slips, you can get quite decent yields as they run around under the trees as a ground cover. It is surprisingly attractive as well. The key lime in front is a must-have in South Florida. It just is. Though putting a plastic flamingo at the base of it is optional.

On to the back yard. Cassava is a survival mainstay and is perfect for South Florida. The thirty plants in the illustration, if well-tended, should give you about 300 lbs. of cassava in a year. If interplanted with bush beans, you'll get an additional five-gallon bucket or so of green beans or a gallon or two of dried beans along with it.

Your main vegetable garden beds can be used both for calorie and nutrition crops. In fall, winter and early spring, beds can be pressed into service for cool-season staples like cabbage and turnips. Daikon radishes, collards, lettuce, kale, and mustard are other good choices. It's very easy to grow all the salad greens you desire by dedicating a single bed in fall to the production of a range of greens. Go to the seed rack and buy some salad green mixes, along with radishes, lettuces, spinaches and whatever else looks good to you. Prepare a nice seed bed and sprinkle seeds all across it, then rake in the seeds and water. Soon you'll have a gigantic bed of salad greens and radishes.

Through the spring and summer your main beds can be cropped with black-eyed peas and okra, as well as sweet potatoes and grain corn. In South Florida, you can also grow most warm-season crops right through the winter, though the cool-season crops won't handle the summer heat. Be sure to plant even most of your warm season crops early in the year—January to February—or the heat, humidity, pounding rain and pests of summer will reduce your yields. Note that I dedicated a big bed to sweet potatoes outside the main gardens. After cassava, sweet potatoes are a top survival crop that will help keep you full. The bed is 9' x 14', which can easily net you 150 lbs. of sweet potatoes.

Your garden beds are only 368 square feet of gardening

space. At least half of this can be devoted to calorie crops and you'll still have plenty of space for more nutrient-dense nutrition crops. You'll be surprised how many greens you can get from that amount of space.

Around the edges of the backyard are multiple trellises which will keep you absolutely loaded down with beans, cucumbers, passionfruit, malabar spinach, chayote squash, true yams, and whatever other climbers you care to grow on them. The vertical space along a fence can be fantastically productive and it's easy to run some strings or wire down the side of a wooden privacy fence. I also recommend a patch in a corner for the excellent Seminole pumpkin. Hemming them in will keep the vines from running over everything.

In the "mixed perennials" and "mixed hedge" area of this plan you can go crazy. Pineapple guavas, Simpson's stoppers, mulberries, coffee, Mysore raspberries (which are truly tropical!), katuk, chaya, edible-leaf hibiscus, cattley guava, Surinam cherries, neem, kumquats, limeberry, clumping bamboo, monstera, gingers, pineapples, grenadilla, hot peppers, various *Pereskia* spp., cocoplum, sea grapes, moringa and more. The possibilities for mixed hedges are incredible. Consider it a 2D food forest—and do not neglect pruning as needed.

Note that I also include a two-bin compost pile and a wider path leading to it from the house. Each bin is 4' x 4' which will handle all your garden waste along with kitchen scraps and hedge prunings. If you want faster compost, do not add hedge prunings. Instead, I recommend making another pile somewhere in the yard for the woodier stuff and letting it rot down over a year or two by itself. Alternately, you can just chop up the prunings and use them for a rough mulch around your fruit trees. Near the compost is a moringa tree, which can be kept cropped for leaves and for "green" material for the compost pile.

A water tank off the roof is a very good idea. If you can get an 800 gallon tank it will serve you quite well—and a good rain will fill it up! A roof gathers a lot more water than you might think. Rainwater is better than city water by a long shot as it does not contain chlorine and fluoride. Garden yields are higher when chlorine is absent.

Coconut palms can be considered a survival gardening staple in South Florida. The nuts are highly nutritious and filling when fully ripe and can be processed to make coconut oil and coconut milk. The coconut water in immature nuts is excellent for your health and can stand in for lunch when you're working in the garden. I find that it fills me up and reinvigorates me as I farm.

Bananas and plantains are reliable producers of fruit if they are watered and fed heavily, hence my placing them beside the house. If you can clandestinely run a washing machine drain out into a couple clumps of bananas on one side of the house and a shower or sink drain out into the bananas on the other side, they'll get all the water they need. The more grey water you can send outside into bananas, the better. It allows you to get two uses from the water and it will also grow you lots and lots of bananas. Getting 40-80lbs of bananas from one stand of bananas per year is not hard. This design should net you around 200-400lbs per year if you can use greywater. If you want to grow even more bananas, you can intercrop them with sweet potatoes in the back yard or take out a garden bed for a few more clumps. Veggies can be grown around their bases to a limited extent. Seminole pumpkins are a good bet. However, if bananas are growing in the yard without greywater, you will have to give them lots of supplementary water. Bananas are lousy producers otherwise.

If you can also grow a jaboticaba, do it. They can produce as many as five crops a year. If you irrigate and mulch it, it will thrive and grow to maturity much faster than if you just let the rain water it.

Finally, Todd and I discussed adding orange trees to these plans, but I have decided not to include them due to how disease-prone citrus has become in Florida. Greening often takes out citrus trees, so you are unfortunately better off skipping them. Except for that key lime. Because you need a key lime.

## Annual yields

Here are some estimates on yields. This is quite a rough guess on my part, but is based on my own experiences. If you feed and care for your gardens and grow high-yield crops, you can beat it easily.

Acerola cherry: 15 lb

Bananas: 300 lb

Cassava: 300 lb

Coconuts: 500 lb

Front-yard yams: 100 lb

Guava: 30 lb

Herbs: 10 lb

Jabuticaba: 20 lb

Mango: 150 lb

Mixed hedges: 350 lb (could easily range from 100-500 lbs., depending on species planted)
Perennial cucumbers: 25 lb

Pineapples: 20 lb

Seminole pumpkin: 50 lb

Starfruit: 150 lb

Sweet potato bed and front-yard sweet potatoes: 200 lb  Various trellised crops: 200 lb

Vegetable gardens: 500 lb

Yams 75 lb

## Total yield: 2,995 lbs

And this is on 1/10 acre! Aren't you glad you're in Florida? Tropical crops can be marvelously productive and we can grow year-round gardens. Plus, we have the best beaches, though that has nothing to do with survival. Unless you count our need for Vitamin D.

This little yard plan is just to give you some ideas. It assumes a South Florida, USDA zone 10/11 climate that allows coconuts and mangoes and other tropical plants.

CHAPTER 9:

# Dealing with Pests

So, is it time to talk about iguanas yet?

No, not yet. We'll cover them in the next chapter. In this chapter, I'll talk about some of the pest issues you are likely to face in South Florida and how you can deal with them in a reasonable manner without blasting your yard with malathion, napalm and/or DDT.

Instead of giving you a full list of pests and a prescription for murdering each one of them, I am going to look at the garden holistically.

## Plant Low-Care Plants

Some vegetables don't seem to get eaten by much. My lettuces sometimes attract aphids, my yard-long beans attract fire ants and aphids, my cabbages attract cabbage worms, my summer squash attracts vine-borers, and my tomatoes attract

every pest known to man—but other crops don't get eaten by much of anything. True yams are consistent growers and remain untouched by insects. Cassava does excellently with no pest control. Chaya almost never gets eaten by anything. Sweet potatoes usually make potatoes without complaining much about the occasional bug holes in their leaves. Seminole pumpkins often sail through vine borer attacks. Bugs ignore my jicama and chayote. Everglades tomatoes don't seem to be worried much by hornworms. Starfruit and bananas rarely have problems. There are definitely species that don't seem particularly vulnerable, either because they make pest-repelling compounds, or outgrow issues, or just don't have many predators in South Florida. That isn't to say you won't discover a problem in the future with one of those species, but they definitely do better than most.

When you plant your gardens, plant multiple varieties of vegetables and trees. You'll soon discover which ones handle pests better and you can let the system fail its way to success. Seriously—things dying will teach you better than trying to keep everything alive. There is a time to plant a bunch of plants and let nature take its course, teaching you which varieties are suited to your yard and which aren't. I based *Totally Crazy Easy Florida Gardening* on this concept and it works!

## Base Fitness Levels

Happy, well-watered and tended plants are much more pest resistant than plants that are struggling and undernourished. It's a common enough observation that many gardeners have seen it. You have one kale plant that is rich, green and barely touched by insects, and another one that is yellow, sickly and absolutely chewed up by insects. We might assume that the insects caused the latter plant to be sickly and yellow, but the opposite may be the case. There is evidence that unhealthy plants attract insects which then chew them up and bring on the death of the plant, recycling them into the ground. Therefore, your primary goal should be to grow very healthy plants right from the beginning. Compost, liquid fertilizers, enough water—all these things help stave off pests or allow the plant to shrug off damage. Imagine two people: one of them fit and healthy and living on a diet of grass-fed beef, farm eggs, organic vegetables, water, and live fermented foods; the other obese and diabetic, living on a diet of potato chips, TV dinners, beer, soda and boxed macaroni and cheese. Both people are exposed to the same pathogen. Which one do you think will fight it off better?

Some years ago my grandpop was pruning one of the trees in the backyard of his Fort Lauderdale home. He was in his 70s, and was standing on the roof of his house and reaching out with a saw to cut limbs. Unfortunately, he overbalanced and ended up falling about ten feet to the ground, breaking his pelvis. Unable to move, he called to the neighbors but was not heard for a couple of hours. Finally, he was found and transported to the hospital.

There he spent many hard weeks recovering, but he did recover. His doctor told him he was "lucky" to have survived such a hard injury at his age.

"If it wasn't for your level of fitness, this probably would have killed you," the doctor said. Unlike most men his age, my Grandpop maintained a high level of fitness, often biking over a hundred miles per week. He would bike down Alligator

Alley, he would bike to Miami, he would bike along Fort Lauderdale beach—he was always moving. Because of that practice, when he broke his pelvis his body was able to survive the injury and heal. If he had broken his pelvis as a 300lb, sedentary senior, he probably would not have lived another month, let alone the extra decade he survived.

Maintain your plants at a solid base level of fitness and they do better with pest injury. They may even produce compounds that repel pests, sending the bugs off to eat your neighbor's sickly virgin lettuce instead of your Chad dinosaur kale.

Micronutrients play into this. A fertilizer like 10-10-10 may make things grow and look green, but it does not meet all the requirements for complete plant nutrition. Imagine eating pure fat, pure protein, and a pure carb without any other nutrients; perhaps a diet of leaf lard, whey protein and white sugar. Yes, you get your "macros," but not your "micros!"

Fish emulsion, Steve Solomon's Mix, DynaGro, Jack's Plant Food, compost, kelp meal, ashes, coffee grounds, seaweed, eggshells, etc., can all add micronutrients to your plants. If a nutrient is missing, plants may attract pests.

Yet—after all this—sometimes plants will still be destroyed. We don't always know why pests attack certain plants or cause a massive amount of damage one year while skipping your garden in the next. Keeping plants healthy is a good start, but there's more to the picture.

## Mix Everything Up

This ties into another one of my gardening techniques, which I call, in precise scientific language, "making a huge mess."

A mess of plants is a haven for predators and a confusing place for pests. I don't deal much with pest issues in my gardens because the complexity of the ecosystem tends to deal with invaders. When you mix flowers and herbs, perennials and annuals, vegetables and fruits and roots and ornamentals all together, you make lots of space for lizards and ladybugs and predatory wasps and other creatures that help control pests.

The only time I have used insecticides in my adult gardening life is when I was very concerned about my food supply and growing gardens on new land without a varied ecosystem. If I felt that an aphid attack was going to keep my family from having vegetables, then I dealt with it. But spraying should be the last option, not the first. It's very convenient and morbidly satisfying to blast pests in the garden and watch them die, but it's not great for you or for your garden. Do you want to eat pesticides? Do you want to kill bees and ladybugs and orb weavers? The fast fix isn't the best fix. Filling an ecosystem with a wide variety of plants lowers pest issues. Also, even if some things die—who cares? You have lots of backups. One year may be great for sweet potatoes, another for pumpkins. The mango might bear a bumper crop one year and need to be treated for fungus the next (copper sulfate works great). Your plantains may love all the rain one season and fruit like crazy, then struggle through the next year's drought. If you plant a lot of different food crops, you will have something to eat no matter what. Redundancy is your ally.

My recommendation on pests is threefold, then. First, base your main gardens on plants that shrug off pest issues and are well-adapted to South Florida. Try lots of different varieties and see how they do, keeping the plants that do great and re-testing those that do not until you are sure they don't like your garden. Second, feed and water well, not forgetting the micronutrients. A high level of base fitness helps plants live through attacks. Third, make a big mess, planting lots of different plants to make an ecosystem that is resilient and harbors a good range of predatory species.

## But I Have to Kill Things Now!

Okay, I get it. Sevin isn't super toxic and works for most insects. The water-soluble form can be sprayed in the evening so it sticks on the plants overnight and has maximum effect on pests. Sevin dust is also good for killing fire ant hills. Amdro fire ant bait works even better on fire ants, though. And don't believe the stories about corn grits killing ants by blowing up in their stomachs or something. It doesn't work. But works well for killing caterpillars and it's just a bacterium. As for the broad-spectrum nasty sprays like Malathion, I just can't bring myself to use them. I can't stand the smell or the thought of eating vegetables sprayed with something "Proven to Kill 50+ Species in SECONDS!"

That said, Malathion is a great name for a metal band.

Finally, a good way to wipe out larger pests like stinkbugs and leaf-footed bugs is to walk around the garden with a bowl of sudsy water and knock insects into it. They'll quickly drown as the soap breaks the surface tension and floods their spiracles, keeping them from being able to breathe. A little dish soap in water is all you need to make an insect death pool. If you have a smaller garden, this method of "knock off and drown" is surprisingly effective and does not introduce poisons into your garden.

But all that aside—I prefer to just plant a bunch of plants, feed them well, and let the weak ones die. Nature does it all the time.

Okay—now—are you ready to talk about the worst pest in South Florida gardens? I think it's time to face ...

CHAPTER 10:

# The Dinosaur in the Hibiscus

One of the most common questions I get from South Florida gardeners is "HELP! THE IGUANAS EAT EVERYTHING I PLANT!!!1!11!!1!" They are the elephant in the room, or, more appropriately, the dinosaur in the hibiscus.

I've seen the damage. They are a real pest, increasing in population and destructiveness as you near bodies of water. Over the years, the iguana population has exploded to horrifying levels. When they first appeared in my grandpop's backyard, they were amusing. We thought it was pretty neat to see an iguana that wasn't in the pet store. But now there are gangs of them hanging out on the seawall and suntanning by the pool. Yet instead of eating insects and being beneficial additions to the garden like most lizards, iguanas are voracious plant-eaters, taking all the crops you planted for your own use.

I freely admit to not being an expert on iguana control, so

I found someone who was. A professional iguana hunter, who stalks expensive neighborhoods at night as a mercenary, taking out these monsters by the dozen.

To round out my lack of knowledge on iguanas and to give readers of this book an idea of how they operate and how they can be controlled, I interviewed Rodney Irwin, known as the Iguana Expert, who is truly knowledgeable as well as being the most entertaining person I have ever interviewed. We hit it off immediately and I am very pleased to be able to record our conversation here.

**David The Good:** I'm writing a book on South Florida gardening. As I was working on it, I thought that the number one thing that I always hear from gardeners is what do I do with these iguanas? Well, I haven't dealt with the iguanas a whole lot because most of my gardening down in South Florida, I was younger and the iguanas have gotten to be a massive plague since then. And so, as I was working on this book, I said, you know what? I really need to call an iguana expert. And you are a professional iguana hunter. Tell me a little bit about what you do and how you got into it.

**Rodney Irwin:** In essence, I am an invasive non-native reptile removal specialist. And all that means is I try to protect the native species we have here from invasion by invasives from other countries. And I started doing tegus and I still do tegus *[ed. note: tegus are a giant invasive carnivorous lizard from South*

*America]*. And tegus are a bigger problem even than the iguanas because they eat our native species. And it's a huge problem. It's not as sexy as pythons, so it doesn't get the press that it needs, but it's starting to become something that people understand and know about now.

**David The Good:** How did the iguanas first end up in the United States?

**Rodney Irwin:** Well, they've been here for a long time. It's just, the population has now exploded. I mean, there's a tipping point when, "Oh, there's an iguana, that's pretty." "Oh, that's unique." "Oh, that's different." To, "Ah there's 10 of them on our dock, crapping on our boat. And I stepped in some this morning and that's a big problem." That is part of the problem with iguanas. The other part is they have to have a specific vitamin and mineral load in order for them to function and survive. And the types of plants we have here in South Florida are not heavy with those materials. So, what they do to make up for that is they eat a lot and it comes out the other end, they get what they need to survive, but it's kind of ugly.

**David The Good:** It's rather like in *Jurassic Park* where they were engineering the dinosaurs so they couldn't go without lysine. And then they find out at the end that some of the dinosaurs have actually escaped and that they're eating specific crops high in lysine.

**Rodney Irwin:** Exactly. I can tell you here, number one is bougainvillea. I could show you bougainvillea plants, that before I started taking the iguanas out, there were sticks. There wasn't a bloom on them, and now they're just punched out with a bright red bloom and they look like bougainvillea should. And if I don't stay at that, you'll see the bloom dissipates more and more and more. Other than bougainvillea, they love hibiscus, they'll eat anything that fits their palette. And I don't know how they decide, which is which, but they tend to like soft fleshy leaves and colors.

**David The Good:** I've been told by a lot of gardeners who have tried to put in fruit trees and edible vegetables in their backyards that they can barely get things to grow, because the iguanas find them every time. They can't babysit them all day long. And no matter what they do, the iguanas come back and wolf down all of their food.

**Rodney Irwin:** Yes. I can't really talk much on that because here in South Florida, where I live, we don't plant a lot of things that are a food source, like potatoes or whatever. I mean, we do, but they don't seem to eat them.

**David The Good:** There are a few edible hibiscus family members that we like to eat the leaves on. And some of my gardening friends were like, "Oh, I can't grow those edible leaf hibiscus. They're just stripped down to sticks." And I said,

"Well, I know those iguanas love hibiscus. I'm not surprised they like those huge soft leaf types that you would make a salad out of."

**Rodney Irwin:** Geography comes into play here. The people that you're talking about, where do they live?

**David The Good:** Some of these gardeners are in the River-land area along the canals in Fort Lauderdale, which is near where I grew up and where my in-laws live. The *Jungle Queen* travels along the New River there.

**Rodney Irwin:** That part of South Florida has similar weather to what we have down here in Homestead. If you go farther north, then you'll see a sharp decline in iguana issues. But any-time you have, let's say an exclusive resort or a beach, high-end beach area, they plant soft-tissue, pretty plants that bloom. And that causes the iguana population to continue and grow because they've got a lot of food.

**David The Good:** That makes sense. Just like the deer popu-lation farther north grows during a really good year for plants.

**Rodney Irwin:** Right. Most of the iguana population from South Florida to Key West, definitely. The Keys are eaten up with them, and then northward up to Delray. I don't know how far inland they go, but they like to be near water, though they're not aquatic.

**David The Good:** I've noticed that iguanas retreat to the water. If you walk up on them and they're out on the dock, they'll hit the water and they swim like fish. Beautiful.

**Rodney Irwin:** That is their primary defense. They run right to the water. I think they can run to the water when they're dead. I swear I've shot them in a tree three or four times, and they fall out of the tree and sit there for a second and just zoom straight to the water.

**David The Good:** If you want a garden and you live near a canal, what would be the best way to keep them from coming and eating everything that you plant?

**Rodney Irwin:** That would be impossible. There are companies down here that advertise that they'll protect your house from iguanas. But what they're really doing is they're really fence companies and they have this iguana-proof fencing. So, if you want to protect your stuff, all you got to do is surround your entire property and your swimming pool with a five-foot-tall iguana-proof fence. And then every tree in your yard has to have it. And the reason for that being is they sleep in the trees. They never sleep on the ground and they'll go up a tree to the highest branch that's big enough to support them, wrap themselves around it, close their eyes, and sleep like the dead. Many people see iguanas in the day and think, "Oh yeah, we've got an iguana right over there." But you go back there at night and there's no iguana anywhere. They're all in the trees.

**David The Good:** It sounds a lot like chickens. Chickens will go up to roost at night, which gets them away from predators. They go up as high as they can. And they sleep tucked up in the branches. And when they go into their sleep cycle, they're very, very dumb. Are iguana the same way?

**Rodney Irwin:** Yep, exactly. They go up to a point where if anything climbs the tree and goes after them, they'll feel it when it comes onto the branch that they're on. And they just let go.

**David The Good:** They just let go and drop?

**Rodney Irwin:** Yes—and you think they might injure themselves? No. Iguanas can drop 30, 40 feet onto concrete and they'll get up and run away. That's something that's in the news every time we have a good cold front down here in South Florida. "They were dropping out of the trees." And they sell it like they fall out of the trees and die.

**David The Good:** No, you just bring them into your house and they warm up again and they run away.

**Rodney Irwin:** Yeah. But you don't have to. Eventually they will warm up on their own and go right back to it. And on cold days, they're so sluggish they can hardly move. I mean, physically, even the ones that are down low. I see a lot of the smaller ones in the winter. And you can walk up, just pull them right off the leaf, hold them in your hand. And their body temperature is such that they can't move their muscles and make things happen for a while.

**David The Good:** If you were going to try and wipe out the iguanas that are in your yard, it seems like the winter might be the time to get them?

**Rodney Irwin:** Yes. And night is the time to get them. In daylight, they've got incredible eyesight and hearing, and it's really difficult to catch them in the daylight, catch them by hand or even with a gun. But at night, especially when it's cold, they'll be up the top of the tree. And I shoot holes in them until they fall down.

**David The Good:** You could just go and shine a flashlight into the trees in your yard and find them up there?

**Rodney Irwin:** Yes, sir.

**David The Good:** I never really thought about what do they do at night, because they always seemed to be down there in the bushes and right in the backyard and right along the canal and sunning themselves. I thought they just tucked themselves in the bushes or something. So, that would be the way to do it instead of trying to chase them all over the terrain during the daytime and run and then worry that you're pointing your air rifle at your neighbor's Mercedes.

**Rodney Irwin:** Exactly.

**David The Good:** Look up in the trees.

**Rodney Irwin:** Yep. The places that I hunt, it would be a very bad thing. If that pellet goes somewhere other than where I told it to go, and by hunting at night and shooting upward into the trees, that's not an issue. There are people that shoot them in the daytime on the ground and they pay a price for it occasionally. But they are darn fast and have such good predator sense. It's hard to get close enough to them with a pellet gun to shoot them when they're on the land in the open.

**David The Good:** I tried stalking some in my in-law's backyard because they said, "Those things keep eating this and they keep eating this." And I said, "Oh, I'll take the pellet gun out there and try." Man alive—they're fast! I was probably 40 feet away and they started racing down the fence and over into the neighbor's yard.

**Rodney Irwin:** Yep. And in terms of how to protect your property, the people that sell the iguana-proof fencing, even if it worked the iguanas would just go up to the trees and to your neighbor's trees and from there onto your property or onto your trees. And how do you cover your gate with iguana proofing? Realistically, the only way that you can make a dent in the iguana population coming into your yard is to shoot them.

**David The Good:** I know that somebody's going to say, "Well, can't you do anything? Can't you like put wolf urine out there or some crazy thing?" They're always telling me this stuff with deer, "Oh I got to keep deer out of my garden." Hardly anything works to keep deer out of the garden other than hunting them or having a fence that is massive. And so, I know some people will say, "You don't want to kill those! You don't want to kill them. That's so violent!"

**Rodney Irwin:** Yep.

**David The Good:** But it's not like the population has gone down.

**Rodney Irwin:** Oh no. They originally came into the Keys, probably on boats from the Caribbean bringing fruit. And it doesn't take them long to establish a population where they can rapidly increase from there. I'm at right about 3,000 that I've taken out so far.

**David The Good:** And I would imagine if you're going to count how many females were in there, each of those females is capable of producing many, many more offspring. So by killing 3,000 iguanas you may have saved people from 10 million iguanas over the next decade or so.

**Rodney Irwin:** Yeah, that's part of my sales presentation. One will hatch out 50 to 60 per year, and most of the time they're in places that do not have enough of a predator load to keep the population in check. This resort where I've been working, they have a thousand cats there that are fed and counted every day. This lady's cat saved her life in her house in Connecticut when her house was on fire. She left millions of dollars for the cats at Ocean Reef. The cats at Ocean Reef are so well fed, they wouldn't stand up and think about chasing a lizard for food. And by and large, there's just not that much, that will take iguanas out as fast as they regenerate.

**David The Good:** South Florida is a very, very tame environment because it's all separated into yards and people have done things like spray the fire ants. You get a little bit out of the city and the fire ants are sometimes incredible. But growing up, I mean, we had a fire ant hill here and there, but almost everybody killed them out. But when I went up to Polk County and stayed with some relatives there, the sheer amount of fire ants out in that sandy yard where nobody did anything was incredible because it was not a tame environment. And if you keep chickens in your backyard in South Florida, nothing's going to kill them unless the neighbor's dog comes in. Yet if you try to keep them in the Ocala National Forest, you have maybe 15 different animals that like to eat chicken. I imagine part of the problem is we made a really tame environment, and then we allowed in a creature that eats all the nice lush landscaping plants that we planted. They basically live in the iguana Garden of Eden.

**Rodney Irwin:** Yep. And it's the same thing for the tegus. Number one on the tegus' food list is eggs and that's what makes them such a problem. They'll go into a nest of threatened, protected American crocodiles. They'll go in there and eat as many as they can cram in their body and destroy the rest of them. So you lose a whole lot of snakes, birds, whatever. Yeah. Tegus are a real problem.

**David The Good:** Now do the tegus eat the iguana eggs?

**Rodney Irwin:** Not that I've ever heard of, but I'm sure they would, they will eat the small iguanas, anything that's of a size that they can put in their mouth and they can run down and catch, they'll eat. And iguanas definitely meet all of that. It's just the iguanas are not really in the tegus geography. Tegus are aquatic. Yeah, they'll eat small rats. They'll eat anything that's on the ground. But yeah, there's not that many iguanas in the tegus' areas.

It's a real interesting lizard. They haven't been here that long and they're just going like crazy now.

**David The Good:** There was a lizard that I remember showing up when I was maybe in my early 20s. So about 20 years ago, I remember for the first time seeing this little fat lizard with a curly tail in a parking lot, the tail curls backwards over its back.

**Rodney Irwin:** Yep. We got those by the plenty. We also have tokay geckos, which are about eight inches long.

**David The Good:** I've heard about those.

**Rodney Irwin:** They make that noise all night long. And they're real fat.

**David The Good:** And they're real fat?

**Rodney Irwin:** Yeah. And they will bite the crap out of you.

**David The Good:** Oh my gosh.

**Rodney Irwin:** So, all things considered, iguanas are friendly to a point. They're not really dangerous. I mean, if you grab a big one, he can tear your arms up with the back feet claw. But other than that, they don't bite. And if they do bite, they're not going to hurt you. The primary problem with them is they eat everything and crap all over everything.

**David The Good:** Right. And for gardeners, it's a major issue. People are pulling their hair out. They want to feed themselves. They want to feel like they've got a little bit of food in the backyard. And then as soon as they start putting stuff in the iguanas come and wolf it down.

**Rodney Irwin:** Yeah. Stopping iguanas from acquiring a food source is just about impossible. And as you were saying, some people might think, "Well, you could capture them and move them to a different location and let them go." And that's against the law. Fish and Game would have a real problem if they caught you doing that. Now it's illegal to possess a live iguana.

**David The Good:** But you're allowed to shoot them?

**Rodney Irwin:** You can beat them with golf clubs. No problem.

**David The Good:**
Sounds like they just gave up. It's like, "All right. Just kill them."

**Rodney Irwin:** That's pretty much it. All of that took place in the last six months or so because they're getting so many calls. People see a four-foot iguana in their backyard, first thing they do is pick up the phone and call Fish and Game.

**David The Good:** Right.

**Rodney Irwin:** "You just send somebody out and get rid of this evil, giant lizard in my backyard." Well, I'd say don't hold your breath. But they do get a lot of that.

**David The Good:** When I was down on the island of Grenada, I was out one day with my friend Moses and he had a little sawmill and was cutting up some mahogany. He said, "You got to come and see this tree. You got to film me cutting this tree up. It's a big, big tree." So, I went down and I took my camera and I did a little filming. And on the way back, he said, "You want to go get some rum and something to eat?" I said, "Yeah, sure." We stopped by these folks that were on a hillside and it didn't look like any restaurant I'd ever seen. They had a tent out there, they were playing dominoes and they were drinking rum and smoking marijuana. And he comes down and he goes, "Hey, hey, hey." And they're like, "Hey, hey, hey." And he says, "Sit down. I'm going to see if they have some food." And he asked the guy and the guy says, "Yeah, yeah." And he opens up this pot that he's got over a fire and he's cooked an iguana in there. And he said, "You want some?" And I said, "Sure, why not?" I was hungry and never tried an iguana before. And it was really delicious meat. On the island we rarely saw a big one, because even in the rich neighborhoods some Rasta guy would come by and he'd be shooting iguanas and taking them down to the market and selling them. If they didn't eat them themselves, that is. They were constantly eating them. Have you ever tried them with all the hunting that you've done?

**Rodney Irwin:** I have. My housekeeper was from Honduras and she cooked them. The eggs are supposed to be highly prized and I didn't care for the eggs at all. And the meat, I've eaten so many different kinds of meat. It wasn't something that really blew my skirt up. And I get that line of thought quite a bit. People ask, "What do you do with them when you shoot them?" And I tell them I feed them to the crocodiles at the crocodile sanctuary. And they say "Well, people eat them." Yeah. Try to find somebody to eat 50 pounds of them in one night. I mean, there are people that eat them, but compared to the amount of iguanas here in the state, there's no value to iguana meat.

**David The Good:** That makes a lot of sense, because on the island there was a population that ate them regularly and was used to eating them. There was a demand for them there, but there's simply not much demand for something that weird. It's like, "I don't want to eat a lizard. I want to eat chicken." If you had wild chickens, maybe somebody would eat them but a cold-blooded, gross-looking freaky lizard with spines on it, I guess most people just don't have a taste for it.

**Rodney Irwin:** Yeah. There are people from Central and South America here, but even if they all started eating the iguanas tonight, it wouldn't even come close to the amount of iguana meat that's out there.

**David The Good:** I suppose iguanas could be your emergency meat supply if civilization collapsed. And you could dig trenches in your garden and turn them into compost. But I don't think you could make iguana skin boots or anything like that.

**Rodney Irwin:** Nope. I wish there was. I really wish there was. Best thing is to put them in the alligator and crocodile ponds.

**David The Good:** When do iguanas nest? Is there a season? Could we could find and destroy their nests?

**Rodney Irwin:** Yes, they are seasonal and they dig in the sand or in the dirt or any place where it's something other than solid coral rock. They'll dig a tunnel in there and lay the eggs and keep and protect the burrow. As soon as the young hatch out, they never go back. And eventually that tunnel and hole will collapse, which is causing a lot of problems. Water Management's having a lot of that problem. An iguana burrow will collapse and part of the bank will slide into the canal and Water Management, they're all about making sure that they can dump water from one place to another, if there's a hurricane coming or something. But if you've got a canal with a bank that's slid into it over and over again, they can't have that. They have to go in and dig it out and replace the fill.

**David The Good:** Any place with loose dirt or sand is a potential nesting area?

**Rodney Irwin:** Yep.

**David The Good:** And they dig extensive tunnels?

**Rodney Irwin:** Oh yeah. I have a friend who works for Water Management. He said someplace he was working last year there were tunnels 50 feet long, and there were intersections and they were crossing over each other. And some time after the eggs hatched, the tunnels finally collapsed and he said it looked like a freeway system.

**David The Good:** Are there multiple iguanas digging tunnels that are intersecting with each other or is it just one very ambitious individual?

**Rodney Irwin:** Oh no. That was a place where there was a lot of iguanas. The canal system here in South Florida probably has the highest iguana population on earth.

**David The Good:** Yikes.

**Rodney Irwin:** You've got canals with people living on them with soft rock and sand. People's backyards go right up to the canals and they all have plants and pretty, pretty stuff growing. And when they dig a canal, most of the time they pile up the dirt they just dug and put it in a big mound that runs parallels to the canal. It doesn't get any better than that for iguanas.

**David The Good:** When is their nesting season?

**Rodney Irwin:** That would be in late springtime. I got some great pictures. I had pictures of four of them I shot, four females. And I opened them up and every single one of them was just full of eggs ... you couldn't put another egg in that body.

**David The Good:** That was another thing when we were down in Grenada, somebody had pulled the eggs out of an iguana and they were eating them, which was pretty grody to me. I did try one just because I write for a living. It was kind of like a weird yellowy paste. I did not like it. I'm with you on that. They must be prolific breeders, then.

**Rodney Irwin:** You're talking about an animal that lays, I'd say average 45 to 55 at a time and that's a lot of hatchlings and the hatchlings are really fast. Their eyesight and their ability to move is almost as good as the full grown adults.

**David The Good:** I've seen them before. They're a lot like the Cuban anoles where they move real quick. And from what I've read, I think they eat insects when young and become more vegetarian as they get older.

**Rodney Irwin:** Yeah. Insects would be a very minor part of their diet, though.

**David The Good:** Are they full vegetarians when they're older?

**Rodney Irwin:** Yes. I won't say a hundred percent, but actually I trap tegu, you know, live capture trap, and I have caught iguanas in my tegu trap, but that's really rare. But it goes to show you that they will go into a trap with egg as bait.

**David The Good:** It may be a little opportunistic.

**Rodney Irwin:** Yeah. I suppose I trap in a lot of different terrain and I've caught iguanas mainly closer to the water.

**David The Good:** I remember you telling me when we talked before this interview that you go hunting in an awesome off-road vehicle and use a 30,000 psi air rifle. I mean, it sounds like a dream job.

**Rodney Irwin:** Yeah. I get that a lot.

**David The Good:** Like people will say, "How do I do that? Why am I a cashier? I could be shooting things in trees at night!"

**Rodney Irwin:** When it's August and the mosquitoes are really bad and you're in a lightning storm and it's four o'clock in the morning, being a cashier starts looking pretty good.

**David The Good:** I could see it. There's always something.

**Rodney Irwin:** After I shoot them I have to physically get them. And a lot of times that requires walking into a lot of stuff where I can't even see my feet.

**David The Good:** I imagine that's actually kind of dangerous when you have alligators and who knows what else?

**Rodney Irwin:** Yep. Snakes, scorpions, you name it. But if I don't get the body, I don't get paid.

**David The Good:** It's a bounty.

**Rodney Irwin:** Yeah. I'm a hired gun.

**David The Good:** We've talked a lot about iguanas and of course iguanas are on everybody's mind as they've become such a massive problem. And you mentioned tegus being the next problem. Is there another crazy threat you see South Florida facing, it looks like it's just been successive invasions over time. What do you see coming that worries you at this point?

**Rodney Irwin:** We have 500 invasives in Florida. Over 500, but that includes invasive plants, fish, reptiles, everything. The only thing that I really I would have to say would be tegus. Tegus are hard to kill. And they live most of the time in places that aren't that populated by people. Down here, we've got a lot of farmland that was farmed in the 30s and the 40s, and they finally figured out it was too low to farm because we'd have real wet summers. And what happened was they cleared a lot—I'm talking 10,000 acres—and it's ideal habitat for tegus.

**David The Good:** And if there are no people around the population just gets ridiculous, because you're not seeing it.

**Rodney Irwin:** Yep. And an acre can only support X number of tegus. And when it reaches that point, they expand and that's what's been happening as we speak, is it's just a matter of food supply and enough ground so they can forage and do well. They're real good at capturing, killing other species that a lot of them are, we don't consider important like rats, but yeah anything small.

**David The Good:** How big can a tegu get?

**Rodney Irwin:** Five foot would be a big one.

**David The Good:** That's a pretty big lizard.

**Rodney Irwin:** Oh yeah. And he will bite the crap out of you. You get into it, even a small tegu a foot long can bite you in a hand and rip your fingernail off. And I said a foot. When you get into two foot, three foot, they'll bite your fingers off. You should see my body. I've got some scars on me. I got bit in the hand and d—near lost my hand. They're like a pit bull, when a tegu bites down, he don't let go.

I was doing a shoot for a French company. And I'm out in the wild. I've got a big female. I'm mic'd up. They got two cameras running. I'm standing there with the female being interviewed and that thing turned and just bit the s— out of me and wouldn't let go. And all these French guys, they're jumping up and down. Now I'm bleeding all over the place. I'm just spurting blood. "What do we do? What do we do?" I said, "My truck, under the seat. Pistol." "Huh?" The whole time I'm standing there with this tegu clamp down on me. And gosh, goes over there, reaches under the seat, comes back with my pistol. He's holding it with two fingers, like it's a rattlesnake about to bite him. And about the time they got to me, it let go. But I'm perfectly willing to blow their head off.

**David The Good:** Sure.

**Rodney Irwin:** When you got one clamp down on you. All you can think is, it's like getting bit by an alligator. You know, this has got to stop, by any means necessary.

**David The Good:** I've heard reports of a lot of stuff moving into the Everglades. Anacondas, piranhas. Even cobras, somebody told me. I mean, there's like weird, weird stuff. It seems like, because I guess everything can live there.

**Rodney Irwin:** Yeah. But I'd have kind of a problem with that. Having said that I caught an anaconda. It was the only one in years and years anybody'd seen and I told Fish and Game and they about had a baby over it. They're afraid the anacondas going to be the next python. But there's so few anacondas that I don't think there's enough of them to let that population grow into something. The one I caught I'm sure was released, because of the place that I caught it and it was perfect. I was hunting tegus with my dog and all of a sudden, my dog just bailed off into the water about a foot deep, but he was running like crazy. I'm thinking, "Well, I guess I better go see where he is running at." And I chased him now and got there, there about an eight, nine-foot python striking at him. And I wasn't sure what it was at first. I'm looking at it thinking, "Well that's not a python. No that's an anaconda."

As for piranhas, and if we had piranhas, it'd be the end of the world down here. If piranhas ever made it into our canal system, that would change a lot of things. And FWC and Water Management, they look for that in a big way. Trying to think of what else in the Everglades shouldn't be there. Not really a lot that's not native, but the Everglades are just so different. You're up there off 41 and it's all the river of grass. And then you come down here and we have a lot of grass land, but we also have cypress heads and we also have pine land that really doesn't flood.

**David The Good:** Yeah. And it can be right next to it. Like there's a pond apple slough, and then you go a little higher up and there's pine land and you go over here and there's something else. It's a varied ecosystem.

**Rodney Irwin:** But as far as new invaders coming on, I don't know. I have no doubt there are invaders on the horizon, but I don't know what they are yet. But they're not at a level that it's become prevalent. And I'm sure as soon as it starts happening, now it's going to go right to the media.

**David The Good:** Yes. And I think probably a lot of people still haven't heard about tegus. That was a new one for me when I was reading your website. "Huh? Wow. These invasives just keep coming."

**Rodney Irwin:** Yes, they do. It's a lot of things like tegus and probably iguanas, this is better territory than their homeland.

**David The Good:** Right. Because they don't have the predators. It's just not there. There's nothing to hold them in check when they get here.

**Rodney Irwin:** Yep. Tegus have an abundant food supply, which unfortunately is our natives.

**David The Good:** That's terrible. As I think about this interview, I may have your portrait drawn by our artist with a couple of terrifying creatures or something.

**Rodney Irwin:** Sure. Story of my life. And you know, iguanas are funny because they look so evil, but in actuality they're one of the least evil reptiles that we have. I mean, they don't ever attack. They won't eat your dog. And if you grab one, he's going to scratch you and you're going to let go, and he is going to run away and even if he does bite you it's not going to hurt.

**David The Good:** I was always rather fond of them. When I was a teenager back when they were still pets, I had a friend whose grandparents lived next door named Mike Schwartz. He had a pet iguana and was maybe a year younger than me. That iguana would sit on his back porch up on a curtain rod. And it would just hang out there all day and he would feed it lettuce and stuff, but it was his pet iguana and he would just take it down and you know, carry it around with him. Really chill animal.

**Rodney Irwin:** They can be like that. But as a rule, they're not a real good pet. Actually, believe it or not, tegus are a better pet. They could do so much more damage, but a tegus is a really smart reptile. Smartest reptile in the world, probably.

**David The Good:** So, it's tamable.

**Rodney Irwin:** Yeah, it comes to your house and it doesn't take him long to figure out, wow, there's no threat here. So, I don't have to be in fight or flight mode and you know, they'll jump up in your lap and they love to watch TV and have you pet him on the head. But in the wild you try to grab one and you're going to regret it.

**David The Good:** That makes a lot of sense. A lot of animals are that way. Where they just have to be handled, et cetera. But don't tell my kids because they're totally going to want pet tegus.

**Rodney Irwin:** Yeah, and people who have them as pets do get bit. But in just about every single case it's because the human did something stupid and they probably did not know they were doing something stupid. Tegus and iguanas are cold-blooded. So, if the tegu has been sitting underneath the heat light, that's set at a hundred degrees for the last hour and you reach in there, go to grab him. He's going to bite you. It's not because he doesn't like you. It's just because when they're hot like that, you make quick movements, they're going to bite first and ask questions later.

**David The Good:** Right. They're totally hopped up.

**Rodney Irwin:** Yeah. But it's pretty much like that with any cold-blooded creature.

**David The Good:** Thank you, Rodney—I really appreciate your input. I think people will find all this information both useful and entertaining.

If you want to learn more about Rodney and his work, or hire him to help with your iguana (or Tegu, python or other terrifying invasive animal) issues, you'll find his website at www.iguanaexpert.com.

\*     \*     \*

Along with my interview of Mr. Irwin, I also asked some gardeners to share what worked for them.

**Stokely Marco writes:** Hey David,

Living on a lake brings with it lots of wildlife. Some of it is welcome, like the wide variety of birds. We even have an osprey that hunts our lake on a regular basis. But the large abundance of iguanas are not welcome. Iguanas love living on or near water, so South Floridians on water will have more problems with them.

## Solution #1

I shoot them with a pellet gun, sometimes composting and other times I just feed the turtles. Have never tried to eat them as my wife would never go for that! Several people on the lake are shooting them, and I have noticed their numbers decreasing, but it only takes one to get through to lose 2-3 months of growth if not the plants altogether.

## Solution #2

My best line of defense are the enclosures I built. This is where I grow a majority of my annual vegetable. I tell people to think of a greenhouse with wire vs. glass panels. Kind of a big chicken coop enclosure of sorts. Funny thing is that after I built this I had the city code enforcement come by to "talk to me about the chickens I had along the lake." We're not allowed to have chickens inside the city limits. Someone had called the city thinking this was for chickens. I explained it was my garden enclosure to keep the iguanas from eating all my veggies. We had a good laugh and that was that. There's the beauty of not living in an HOA. Mine is integrated into my terracing and low so as to not be in the line of sight to the lake. It also protects the garden from other things like squirrels, rats, possums raccoons, etc. We have them all.

## Solution #3

You can put small wire rings around each plant.. They are simple enough to make with wire mesh fencing and a few zip ties. I like the stronger 1/2" type as it is stiffer and holds its shape better. I do this by plant, but I would imagine it would work equally as well for a small grouping. If you really watch the iguanas, they tend to be "drive by grazers," just walking through and eating things at their level. Once you put the metal wire rings round the plants, they don't make the effort to climb up to get to the vegetation.

## Solution #4

Look for or create sanctuary locations on your property. I have a few locations close to the house where the iguanas don't go too much. I guess our presence or our traffic is the deterrent. The real young ones go everywhere but the older ones stay away for the most part. Also taller pots are hard for them to climb so they help protect plants in addition to the location. We observed over time where they ventured on the property. Since our backyard is on a lake, you can imagine that is where most of the encounters take place. But we noticed they tended not to come deeper into the side yards.

We also have vinyl fencing running along the sides with

gates at the front of the house which blocks easy access from the side of the house. In general my side yards are sanctuaries as they are protected on 3 sides by the fence or house. We did the vinyl fence for privacy, but got a secondary bonus of blocking easy access.

I use a combination of in-ground, multi-tier planters I built, towers, and pots to grow in these areas. Whatever holds dirt!

I also put bird netting over my blueberries (in the planters), but more for the birds. Bird netting will block them, but I have seen them get tangled up in it.

## What Iguanas Eat in my Open Gardens

Iguanas do climb plenty to get to other plants growing on fences, but they mainly eat the leaves and some flowers on a few plants. I have not lost any fruit to iguanas. I have mangos, avocados, bananas, lychee, loquats, pineapple and more. But they will strip the leaves off of peppers, beans, squash/pumpkins and passion fruit. I assume they would eat about any young leaf or brassicas. I don't even bother with those in the open. Some edibles that they seem to not eat: cassava, fig, corn, jicama, muscadine, taro, gingers, strawberry and turmeric. They have also not eaten any of my leaf perennials like Okinawa Spinach, basils, cranberry hibiscus and roselle. Interestingly, I have Aji Dolce peppers growing next to pepperoncini, bell and banana peppers and they eat the leaves of all but the Aji Dolce plant.

**Margot Mesnard writes:** Hi David,

In a group of home growers in Sint Maarten, Caribbean, here's what people suggest:

- fence your crops
- wrap an aluminium sheet around trees trunks so they can't climb from the bottom
- trim the branches so they can't jump from the rooftop or other trees
- keep smaller fruits trees low so the branches are too weak for iguanas to climb on
- get lots of dogs and cats on the property to chase them away

**Anonymous writes:** I "follow the science" and "listen to the experts." The scientists agree that iguanas are an invasive species which spreads diseases and has no natural predators. I shoot them with a .22 caliber rifle using Super Calibri cartridges from Aguila. The Colibri cartridges are not "ammunition" because they contain no gun powder. Iguanas are quite intelligent. They avoid my block. It is legal to shoot iguanas on any private property with the permission of the land owner. One neighbor was not enthusiastic about my "hunting" until he went to sit on his toilet and had an iguana looking up at him. Now he texts me when there is one in his yard.

## The Florida Fish and Wildlife
## Conservation Commission notes:

Green iguanas are not native to Florida and are considered an invasive species due to their impacts to native wildlife. Like all nonnative reptile species, green iguanas are not protected in Florida except by anti-cruelty law and can be humanely killed on private property with landowner permission. This species can be captured and humanely killed year-round and without a permit or hunting license on 25 public lands in South Florida.

Green iguanas cause damage to residential and commercial landscape vegetation and are often considered a nuisance by property owners. Iguanas are attracted to trees with foliage or flowers, most fruits (except citrus) and almost any vegetable. Some green iguanas cause damage to infrastructure by digging burrows that erode and collapse sidewalks, foundations, seawalls, berms and canal banks. Green iguanas may also leave droppings on docks, moored boats, seawalls, porches, decks, pool platforms and inside swimming pools. Although primarily herbivores, researchers found the remains of tree snails in the stomachs of green iguanas in Bill Baggs Cape Florida State Park, suggesting that iguanas could present a threat to native and endangered species of tree snails. In Bahia Honda State Park, green iguanas have consumed nickerbean, which is a host plant of the endangered Miami Blue butterfly. As is the case with other reptiles, green iguanas can also transmit the infectious bacterium

Salmonella to humans through contact with water or surfaces contaminated by their feces.

*For deterring without killing iguanas, they suggest:*

If you have an iguana frequenting your area, you can take steps to deter the animal such as modifying the habitat around your home or humanely harassing the animal. Examples of effective habitat modification and harassment include:

- Removing plants that act as attractants
- Filling in holes to discourage burrowing
- Hanging wind chimes or other items that make intermittent noises
- Hanging CDs that have reflective surfaces
- Spraying the animals with water as a deterrent

(https://myfwc.com/wildlifehabitats/profiles/reptiles/green-iguana/)

**Allan Turpin writes:** My pet iguana Abraham traveled cross country with me camping in 23 states along the way. She (her previous owner didn't know she was a female when they named her) was born in Maryland and buried with honors on the shore of the Pacific in California. She was likely the first iguana to camp along the shores of two of the Great Lakes, in Dinosaur National Monument, Yellowstone, Bears

Lodge (aka Devil's Tower), Puget Sound, and at the base of the biggest redwoods... and many other sites including a stop at the Testicle Festival.

**Sue Yellet writes:** My dog has had face-to-face encounters with two that I know of and removed a ten pounder from my food forest. I have placed metallic pinwheels in my garden as a deterrent and bagging papaya fruit.

**Barbara Perry writes:** One was ravaging my garden everyday, so I put a pellet gun on my Lanai in a handy spot. I had covered all of my beds in bird netting to try to keep it out. One day I was on the Lanai and heard some noise so I walked out to the garden. The iguana was trapped In the netting. I ran back to the Lanai and got the pellet gun and shot it in the brain. Then put it in my compost pile. They are invasive, and they are multiplying like crazy here in Southwest Florida.

**Steve Cameron writes:** I once walked up on one in Managua. It saw me and ducked inside a manhole cover.

**Hugh Brackett writes:** I hear they make a nice gumbo.

**Grouchy Old Prepper writes:** Family members in Florida are dealing with them and much of their commentary is unprintable here. Except for lead shot, the only thing that seems to affect them is cold weather (which you guys don't really get much of). On those rare occasions when it does get cold enough, they'll fall out of trees and just lay there. Gives you an opportunity to walk around the yard and practice your golf swing with a sturdy club.

**Free Will Choice writes:** There is a guy in Sunrise (Broward County) named Raj who hunts iguana as a career and eats them too. He has a popular channel "iguana man". He says they destroy the water way infrastructure by making lots of holes to live in off the banks of canals, lakes and bridges which can cause sections to collapse over time. He caught 6 females in a video and pulled close to 200 eggs from them not too long ago. He says they are destroying the eco system down south and the government has an open season for hunting on them like wild hogs and lionfish. He is a great "go to" source for more information. (Raj's YouTube channel is here: https://www.youtube.com/c/TheIguanaHunter)

\*     \*     \*

However you choose to deal with iguanas, they will have to be dealt with in South Florida, especially if you live near water. Good luck and good hunting.

CHAPTER 11:

# Managing Your Harvest

In this final chapter we'll look at what to do with all the food you'll soon be reaping from your Florida backyard.

I must confess: I am not the type to write down recipes or to refer to the recipes of others. For years people have written me saying, "David—it's great that we can grow all these crazy crops you recommend, but how do we eat them? Would you write a recipe book?"

Since I don't like to write or use recipes, I decided to thrust the responsibility of writing a recipe book upon my wife. "Hey Rachel, how about you write a recipe book?" I said.

"I could maybe do that," she said.

"My wife is gonna write a recipe book," I responded to everyone, gleeful that it was no longer in my hands. Rachel never wrote it, though, because she's busy raising children, so it was a short-lived dodge.

Now people write me and ask when my wife is going to write that recipe book I said that she said she was going to write about how to use all the weird tropical crops I'm always explaining how to grow but never how to cook.

I don't know. Maybe one day. Until that happens, I have some suggestions for you on using what you grow.

## Learn to be Adventurous

I met many people from various cultures growing up, so we had plenty of opportunity to try new foods. If you're not afraid to taste-test, there are many wonderful fruits and vegetables out there, as well as dishes that use them. I remember when we first discovered how the flavor of kaffir lime leaves brings out the high note in a curry sauce. Rachel's Aunt Ja—from Thailand—introduced us to the plant. Now we always want one in the garden. I remember when my brother-in-law introduced me to fried green plantains. And when I tried my first miracle fruit with Craig Hepworth, then we ate sour lemons and starfruit and they tasted like candy. These experiences are a lot of fun. Sure, you might not like the taste of some things and may be repulsed by others, but don't be afraid. You'll never know if something is wonderful or terrible if you don't try it. If you taste something interesting from a tropical climate, chances are you can grow it in your South Florida backyard.

## Don't Waste the Extra

Sometimes you end up with a lot more food than you

expect, especially if you've got a huge mango tree. A few options in that case, are:

## 1. Preserve

You can dehydrate, freeze, pickle, ferment and can a wide range of foods. We've turned fruit into alcohol and have turned cucumbers into live-fermented pickles. My grandmother froze piles of mango slices in her deep freezer so we could enjoy mangoes year round. You can make sauces and salsas, and most fruits dry into excellent fruit leathers. Learning these methods of preservation allows you to enjoy the harvest for much longer while keeping produce from going to waste.

## 2. Give Away

One of my favorite things to do with extra fruit and vegetables is to simply give them away. If we've been blessed with an abundance, why not share it? Homegrown bananas, fresh pineapples, sweet potatoes, and starfruit are often appreciated by friends and neighbors who don't know how to grow food or are too busy to do so.

## 3. Sell

A productive garden can also be a source of some side income, especially if you are growing uncommon fruits and vegetables. My friends Chuck and Sarah in South Florida have a large jackfruit tree that bears tons—and I mean tons!—of fruit. After telling me that they couldn't keep up with all the yield, I told them they should consider selling the fruit to a local market. Some time later, they found an outlet and now make money from all the extra fruit. Instead of it being a burden to clean up, now it's a monetary blessing.

## 4. Feed to Animals

If you are fortunate enough to be able to keep livestock, extra produce can be fed to them and turned into eggs, milk or meat. We give a lot of our extra produce to our animals. They eat well and so do we. It's not a waste, as you're taking one form of food and turning it into another. And, let's face it; turning bitter wilting turnips into farm-fresh eggs is a great trade. Some crops can be cooked and fed to chickens, like cassava, yams and sweet potatoes. Chickens are also happy to pick coconut meat out of the shell. Rabbits will eat a wide range of greens, too, and reward you with meat and manure.

## 5. Make Compost

If we have way too much of something, we can afford to be picky and to compost the rest. We'll take the best of our harvests and then throw the sub-par fruits into the compost pile. We usually try not to waste much, but there are times you can get overwhelmed. If you can't deal with 200 lbs. of mangoes in a weekend and you haven't been able to give them away or sell them, throw the worst of them into the compost pile to feed next year's garden.

# Substituting

Though I'm not going to give you any recipes, because I've delegated recipes to my wife, I will tell you something that has helped us a lot with managing our "strange" crops from the tropics. If there's a dish you like, it often works to switch out part of it with something similar that is growing in your backyard.

For example, let's say you want to make mashed potatoes for Thanksgiving. Yet you have lots of winged yams available but had no luck growing Irish potatoes. No problem—just make mashed yams instead. Peel them, boil them, mash them with butter and salt, and just treat them like white potatoes. We did this when a friend came to dinner one evening and he couldn't tell the difference between our mashed yams and regular mashed potatoes. Hunks of peeled green plantains cook nicely in a stew, as does taro root. Moringa leaves are good in chicken soup. Mulberries or black Surinam cherries fill in very nicely for blackberries, blueberries or strawberries.

Here are a few temperate crops and their tropical South Florida equivalents.

# Apples

**Alternative:**

Wax Jambu

# Cucumbers

**Alternatives:**

Ivy gourd, chayote squash

# Irish Potatoes

**Alternatives:**

Boniato, cassava, dasheen, true yams (*Dioscorea spp.*), green bananas and plantains

# Lettuce and Spinach

**Alternatives:**

Amaranth, *Celosia argentea,* chaya, edible-leaf hibiscus (*Abelmoschus manihot*), katuk, longevity spinach, Malabar spinach, moringa, Okinawa spinach, Surinam purslane (*Talinum fruticosum*), sweet potato (leaves), water spinach (*Ipomoea aquatica*)

# Peas and Beans

**Alternatives:**

Black-eyed peas, jack beans, lablab, pigeon peas, snake beans

# Rhubarb and Cranberries

**Alternatives:**

Jamaican sorrel

## Temperate Climate Berries

### Alternatives:

Acerola cherry, cherry of the Rio Grande, jabuticaba, Jamaican cherry, mulberry, Mysore raspberry, *Pereskia* (spp.), Surinam cherry

## Temperate Climate Nuts

### Alternatives:

Chufa, coconut, macadamia nut, Malabar chestnut, monkey pot, tropical almond, water chestnuts, jackfruit (seeds, boiled)

## Turnips and Rutabagas

### Alternatives:

Jicama, water chestnuts

As a final note on one of our favorite substitutes mentioned above, my wife has regularly used Jamaican sorrel calyxes instead of cranberries in cranberry sauce recipes. There's a reason the plant is also known as "Florida cranberry." It's a very similar flavor profile but it grows wonderfully in tropical South Florida, unlike cranberries. I actually think it tastes better. If you aren't scared to experiment, you'll find many tropical substitutions for temperate crops that may not only replace but improve upon the original ingredient in a recipe.

## Conclusion

Though South Florida is a different climate than what is normally covered in gardening books, it's not a bad climate. It's not even a difficult climate. Now that you know how gardening works here, you will have amazing success!

Plant some tropical fruit trees. Plant some cassava and sweet potatoes around them. Fill your yard with delicious and interesting crops from the equatorial and subtropical regions of the world. Learn to improve the sand and feed your plants with liquid fertilizers. Next thing you know, you'll have more food than you ever thought possible—and you'll have it year-round, unlike temperate climate gardeners.

If you want to learn more about Florida gardening and the many amazing edible species that grow in the Sunshine State, I invite you to visit my website at thesurvivalgardener.com and subscribe to my "David The Good" YouTube channel. You'll also find my three other Florida gardening books helpful. It's a wonderful place to garden and I hope this little book has opened your eyes to South Florida's potential.

Now get outside and get sandy—there's stuff to plant!

PRIVATE
BEACH
KEEP OUT!

## About the Author

David The Good is a South Florida native and the author of the bestselling book *Totally Crazy Easy Florida Gardening*, as well as *Florida Survival Gardening* and *Create Your Own Florida Food Forest*. He has launched and maintained productive food forest systems in both North and South Florida, as well as grown a wide range of tropical plants in his annual gardens. David currently lives and gardens near Pensacola with his wife and ten children.

## About the Illustrator

Tom is a professional artist working in pen and ink, watercolors, woodcuts and more. In his free time he enjoys woodworking, banjo-playing and gardening with his wife and children on their Florida homestead. He can be reached for illustration work at tomsensible1@gmail.com.